Legion of Night

Legion of Night
The Underwing Moths

Theodore D. Sargent

Photographs by Harold J. Vermes

Drawings by Katherine A. Doktor Sargent

UNIVERSITY OF MASSACHUSETTS PRESS AMHERST 1976

Library of Congress Catalog Card Number 75-8452

ISBN 0-87023-187-1

Printed in the United States of America

Library of Congress Cataloging in

Publication Data

Sargent, Theodore D. 1936-

Legion of night.

Bibliography: p.

Includes index.

1. Catocala. 2. Insects — North America. I. Title.

QL561.N7S27 595.7'81 75-8452

ISBN 0-87023-187-1

This book is dedicated

to the memory of

PHOCION J. INGRAHAM

He set a young mind free

Contents

Preface

THE UNDERWINGS (*Catocala*) are among the largest and most showy of North American moths. Their size and beauty, and their incredible diversity, have attracted considerable popular attention. They have long been favorites of collectors, and scientists have been fascinated by the challenge of explaining their profusion and variety.

These moths have not been treated in a popular work since Holland's *Moth Book* (1903) and Barnes & McDunnough's *Illustrations of the North American Species of the Genus Catocala* (1918). Both of these works were intended primarily as identification guides, but both have become less useful in that regard, as many changes in nomenclature, and various additions to our fauna (particularly melanic forms), have been made since their publication. These works contained very little information on the life-histories, behavior and ecology of these moths, and neither stressed the interesting biological questions that are posed by such a complex assemblage of closely related species.

The present book is intended to be a popular, up-to-date, and comprehensive treatment of the *Catocala* of eastern North America. It includes (1) a complete survey of the species occurring in that region, (2) a summary of current biological information on those species, and (3) an introduction to the scientific investigations which are being conducted on these moths. I hope that this comprehensive approach will be interesting and useful to both amateur collectors and professional biologists, as well as to persons with more general natural history interests.

A first step in the study of any group of organisms is the correct identification of its members, and an important aim of this book is to provide a guide to the identification of all of the *Catocala* found from Canada to Florida east of the Mississippi River. Toward this end, eight color plates depicting 126 specimens and a series of 71 species accounts are included. The species accounts contain sections on the range, seasonal occurrence, status, larval foodplants, and behaviors of the various species, and so should provide the kinds of information that collectors often desire. Considerations of bait and light collecting (chapters 3 and 4), and a discussion of aberrations and other freakish specimens (chapter 5), should also be of particular interest to collectors. For those persons whose interests extend to rearing and mating studies, the discussion of life-histories (chapter 6) may contain material of value.

Ecologists and evolutionary biologists are particularly impressed with the enormous diversity of the *Catocala*, and a number of suggestions regarding the origin and maintenance of this diversity are included in the chapter on life-histories. The book concludes with a consideration of the functional significance of the forewings (chapter 7) and hindwings (chapter 8) of these moths. Particular emphasis is placed on the potential role of diversity in these structures as a foil to the conservative hunting strategies of predators, particularly birds. These final chapters draw heavily upon my own researches, which have focused on the behavioral adaptations of the *Catocala*.

The results of scientific studies on these moths are often unavailable to collectors, as such results are usually published in relatively inaccessible professional journals. Another important aim of this book, then, is to summarize and interpret these scientific studies, hopefully in a fashion that will encourage and direct the efforts of collectors who wish to contribute to our knowledge of the biology of these moths.

I have not hesitated to include extensive data throughout the book. I feel that such data have great value in providing the facts against which various scientific hypotheses may be tested and upon which future studies may build. I also believe that such data are of interest to many readers, and that the authors of popular books often underestimate the interests and capabilities of their readers in this regard.

This book is based in large part on the observations and experiments I have carried out around my home in Leverett, Massachusetts, over the past eight years. I hope that this emphasis will serve to convince amateur naturalists that important scientific contributions can still be made in the "backyard laboratory," and without the use of complicated and expensive equipment. Patient observation, complete recording of data, and careful execution of simple experiments are the means by which our knowledge of the *Catocala* will most surely be advanced.

I am indebted to many persons for assistance in the preparation of this book. I would particularly like to acknowledge the contributions of the late Sidney A. Hessel, who kindly provided his complete collecting records from Washington, Connecticut, for the period from 1961 to 1973. Charles G. Kellogg provided similarly valuable records from his collecting sites in West Hatfield,

Massachusetts, for the years 1968 to 1973. Additional field data were supplied by a number of my present and former graduate students, including Denis E. Berube, Ronald R. Keiper, and Dale F. Schweitzer. Conversations and correspondence with a number of collectors yielded many valuable facts, and I would like to express particular thanks to John Bauer, Auburn E. Brower, Charles V. Covell, Jr., Richard B. Dominick, Douglas C. Ferguson, Sidney A. Hessel, Charles G. Kellogg, Wayne A. Miller, Mogens C. Nielsen, Charles L. Remington, and Dale F. Schweitzer.

A number of persons supplied specimens for use in the color plates and figures. Among the individuals who loaned specimens from their private collections were Sidney A. Hessel, Charles G. Kellogg, Mogens C. Nielsen, J. B. Paine, Jr., and Dale F. Schweitzer. The following individuals kindly made specimens available from the museum collections under their care. Harry K. Clench (Carnegie Museum, Pittsburgh), Douglas C. Ferguson (United States National Museum, Washington, D.C.), and Charles L. Remington (Peabody Museum of Natural History, Yale University).

The manuscript was read in its entirety by Donald Fairbairn, Denis F. Owen, and Oswald Tippo, and I am grateful to them for many helpful comments and corrections. The errors that remain are, of course, my responsibility.

The staff of the University of Massachusetts Press aided in many ways, and I am particularly grateful for their constant enthusiasm and encouragement.

Publication of this volume was made possible through the generous support of the Bache Fund of the National Academy of Sciences, the University of Massachusetts Foundation, Inc., and individual patrons including David Agricola, Lincoln P. Brower, Robert Dirig, Richard B. Dominick, and Mr. and Mrs. John B. Paine, Jr..

I am happy to express my appreciation to Harold J. Vermes for the skill and dedication with which he carried out the photography for this book. And to my wife, Katherine, for the drawings which add so much to these pages, I extend my deepest gratitude. Finally, to David and Meryl, who nearly made this work impossible — thank you!

Introduction

My introduction to the Underwings occurred when I was a boy of twelve. An enthusiasm for collecting moths led me to the local library, where a helpful librarian suggested that I borrow W. J. Holland's *Moth Book*. I hastened home with my prize, and immediately began to turn the pages, hurrying from one color plate to the next. I savored the elegant hawkmoths, the giant silk moths, the brilliant tiger moths; then, sooner than I wished, I turned to the dreary drab noctuids. But, no! Suddenly, after what was surely a million grayish and brownish nondescripts, came plate after plate of beautiful *Catocala* — magnificent moths, many of great size, and with the most colorful and boldly patterned hindwings. Questions immediately rushed to my mind. Why were these moths so large? Why did they have such spectacular hindwings? And why were there so many kinds? My wonder and curiosity had been permanently aroused.

Many years have intervened, and though I have acquired a proper respect for the grayish and brownish hordes, the *Catocala* have always been my favorites. The wonder that excited my boyhood survived my academic training, and upon completion of the Ph.D. in 1963 (a degree, incidentally, which established my credentials as an ornithologist), I turned to these moths as research subjects. After ten years of intensive study, I am convinced that the North American *Catocala* moths pose some of the most interesting and challenging problems to be found in evolutionary biology. The striking design of somber forewings and boldly patterned and colorful hindwings has been extraordinarily successful, when measured in terms of the incredible diversity of these moths. There are few, if any, comparable assemblages of closely similar species anywhere in the world. Questions regarding the origin and maintenance of such diversity immediately arise, and this book attempts to shed as much light on these questions as current knowledge permits.

The book emphasizes the biology of the *Catocala* moths, and draws heavily on my own researches, which have been primarily behavioral and ecological. My field experience has been confined to New England and the species of that region are dealt with in most detail, but coverage is extended to include some consideration, and illustrations, of most of the species occurring in eastern North America.

This work is in no sense a monographic or taxonomic treatise, though every effort is made to represent accurately the current status of the species and their varieties. I will generally follow the established arrangement of the species as found in the most recent checklist of North American moths (McDunnough, 1938), and will adopt the convention of placing the names of all infrasubspecific variations (forms and aberrations) in parentheses, as opposed to the italics used for species and subspecies names. I am well aware that the naming of forms and aberrations is controversial (see e.g., Masters, 1972; Sevastopulo, 1974), and that such names have no standing in the International Code of Zoological Nomenclature. However, I believe that having names for the distinct polymorphic forms of a species is a great convenience, enabling workers to refer to these forms with single words (e.g., *innubens*, form "scintillans") rather than lengthy descriptive passages (e.g., "the form of *innubens* with a very dark contrasting area between the am and pm lines of the forewing"). Accordingly, I have suggested three new names in this book for melanic forms which have heretofore been nameless. I hope that the convenience of having these names, particularly since melanics are of considerable current interest, will outweigh the disadvantage of adding to the already cumbersome *Catocala* nomenclature.

This book has a scientific emphasis, but it is also intended to capture some of the beauty and romance that are invariably associated with these moths. I have tried to limit the use of technical terms, but have included a glossary which may prove useful on some occasions. Hopefully, many readers will find material of interest in these pages, and perhaps some will be encouraged to carry out studies of their own. Science will be well served if such studies eventually render the contents of this book obsolete.

Every association of moths is with night and mystery and death.

Peattie, 1935, *An Almanac For Moderns*

Of Men and Names I

THE CATOCALA moths may initially attract attention and interest because of their remarkable beauty. The striking contrast of modest forewings and flamboyant hindwings fascinates most viewers. But hardly less intriguing are the incredible common names by which many of the species are known. What great sadness could account for the Inconsolable Underwing (*insolabilis*), the Dejected Underwing (*dejecta*), and the Tearful Underwing (*lacrymosa*)? Might these not be cheered by the company of the Darling Underwing (*cara*), the Sweetheart (*amatrix*), or the Bride (*neogama*)? Better to avoid, perhaps, the Penitent Underwing (*piatrix*) and the Old Maid (*coelebs*). But surely the Serene (*serena*) and Sleepy (*concumbens*) Underwings would make quiet companions.

These sometimes doleful, sometimes romantic, common names have elicited scorn from some recent writers. "I hesitate to perpetuate the continuance of the ridiculous names assigned to this genus, which had better be left to die a natural death," writes P. Villiard (1969). To many, however, such names lend an entirely appropriate aura of mystery and intrigue to these great night-flying insects. Some of those who criticize the common names of the *Catocala* may not realize that they are usually literal translations of the scientific names, and so were intentionally assigned to these moths by the great entomologists who originally described the various species.

Many of the older names relate to sorrow and death (the Widow, *vidua*; the Forsaken Underwing, *relicta*; the Mourning Underwing, *flebilis*), and these names undoubtedly reflect the long-held superstitious belief that moths represent the souls of the dead, flying in darkness and ever seeking light (see Clausen, 1954). Many other names relate to love and marriage (the Betrothed, *innubens*; the Consort, *consors*; the Mother Underwing, *parta*), and these were undoubtedly meant to convey a sense of romantic tragedy. Nearly all of the older *Catocala* names are female, presumably because they seemed appropriate to the men who described these undeniably lovely creatures.

More recent authors have departed considerably from the earlier traditions, though a tendency to use female names has persisted. Herman Strecker, who named many species, expressed the following view:

> I have a sort of old-fashioned respect for the way the fathers of science used to name these things; for instance, the Catocalae all had amatory names, relating to love or marriage, Amatrix, Cara, Relicta, etc., etc. Of course these terms would soon be exhausted, and, in fact, have been; then, names that would in a great measure keep up the connection would naturally be next selected, and the most appropriate ones for the purpose would be those of women famous in ancient history for their lust or talents, or both combined. . . . (1874, Lepid. Rhop. & Het., p. 77)

This practice gained a wide acceptance, for the names of women grace many of our American *Catocala*. From the biblical Apocrypha (*judith*), the Iliad (*briseis*), and the pages of Shakespeare (*titania*); from Russian folklore (*babayaga*), Greek mythology (*andromedae*), and Roman history (*agrippina*); the famous (*sappho*) and the infamous (*herodias*); the virtuous (*irene*) and the not-so-virtuous (*delilah, messalina, cleopatra*) — surely here is lust and talent aplenty!

Another American practice has been to apply the names of entomologists to the *Catocala*. This practice is not without its detractors, and Strecker, again, expressed thoughts on the subject:

> Of upwards of forty species found in Europe and Siberia, none had the names of any scientist, ancient or modern, bestowed upon them. . . . But to us progressive Americans it is owing that the harmony of the Catocala Nomenclature has been broken; Edwards first, with his C. Walshii, and then Grote with C. Clintonii, C. Robinsonii, etc.; it is, however, done, and irrevocably so, and we can only in sadness submit. (1874, Lepid. Rhop. & Het., pp. 77-78)

In spite of Strecker's sadness, the practice has become prevalent, particularly in recent years, with respect to the naming of forms. Thus, we have names such as "hulsti" Reiff, "reiffi" Cassino, "cassinoi" Beutenmüller, "beutenmülleri" Barnes & McDunnough, *mcdunnoughi* Brower, "broweri" Muller, and so on. (These men and others so honored may be identified by referring to their works in the Bibliography.)

One final practice involves the use of descriptive names which allude in some way to a distinctive feature of the moth being described. Examples of this sort include "nigrescens" (the melanic form of *ultronia*), "conspicua" (the form of *ilia* with striking, white reniform spots), and "albomarginata" (the form of *lacrymosa* with pale inner and outer forewing margins).

There has been a tendency to eschew romance in the recent naming of the *Catocala*; while this trend may be applauded by many, it brings to mind these words of Thoreau:

> Color, which is the poet's wealth, is so expensive that most take to mere outline or pencil sketches and become men of science. (1852, *Journal*)

THE LATTER HALF of the nineteenth century was the heyday of naming our North American *Catocala*. Of the 71 species treated in this book, 59 were first described between 1850 and 1900. Six were named in the eighteenth century, and only two have been described in the twentieth century.

Most of the species names with which we will be concerned have been contributed by five men: Augustus Radcliffe Grote (18 by himself, and 2 more in collaboration with Coleman T. Robinson), Achille Guenée (12), Ferdinand Heinrich Herman Strecker (10), William Henry Edwards (7), and Francis Walker (5).

Of these men, A. Guenée in France (1809–1880) and F. Walker in England (1809–1874) had no firsthand experience on the American continent. Nevertheless, they described hundreds of American Lepidoptera from collected specimens which came under their care. Guenée's "Spécies Général des Lépidoptères" (vols. v–x, 1852–1857) and Walker's "List of the Lepidopterous Insects in the Collection of the British Museum" (vols. i–xxxv, 1854–1866) contain the descriptions of those American *Catocala* which still bear their names. Biographies of both men may be found in Essig's "History of Entomology" (1931), and A. R. Grote contributed a commentary on Guenée in *Papilio* (1881).

Augustus R. Grote (1841–1903) was born in England and died in Germany, but spent most of his life in America and is regarded as one of our most important nineteenth-century entomologists. The most active period of Grote's life as an entomologist was spent as curator of the Buffalo (New York) Society of Natural Sciences (1873–1880). He is remembered primarily for his work on noctuids, for he described nearly 1000 species in this family. In more than 600 publications, including many on other families (especially Sphingidae), he covered other aspects of lepidopterology (including larval characters and variation), and various nonentomological subjects as well (particularly the conflict between

Figure 1.1 One of the original plates from Herman Strecker's *Lepidoptera, Rhopaloceres and Heteroceres, Indigenous and Exotic, with Descriptions and Colored Illustrations* (1872–77).

science and religion). An excellent bio-bibliographical essay on Grote has been prepared by Wilkinson (1971*a*).

F. H. Herman Strecker (1836–1901), a colorful contemporary of Grote, was born and died in Pennsylvania. He was both artist and scientist, and is best known to lepidopterists for his "Lepidoptera, Rhopaloceres and Heteroceres, Indigenous and Exotic, with Descriptions and Colored Illustrations," which was published in parts between 1872 and 1877, and which contained fifteen hand-colored lithograph plates, many depicting *Catocala* (fig. 1.1). Strecker's reputation for coveting specimens is legendary, and the story of his habit of pinning desirable specimens from other collections into the crown of his silk hat will no doubt survive as long as lepidopterists reminisce. Strecker's incessant feuding with Grote enlivened the often tedious entomological literature of that era. A brief biography of Strecker has been prepared by J. Remington (1948).

William Henry Edwards (1822–1909) was born in New York and died in West Virginia. He is best known for his monumental, three-volume "Butterflies of North America," published between 1868 and 1897 and illustrated with 152 superb color plates. While few of his 256 scientific papers were devoted to moths, he had a particular interest in the *Catocala*. Perhaps in part because of that interest, W. H. Edwards is occasionally confused with his unrelated contemporary Henry Edwards (1850–1891). Although only one of our eastern *Catocala* species names is attributed to Henry Edwards (*miranda*), he is the author of the names of many western species, as well as of numerous well-known forms and aberrations. His *Catocala* names were always those of female characters from Shakespeare (e.g., "celia," "cordelia," "phrynia," "nerissa"). Brief biographies of W. H. Edwards and Henry Edwards have been prepared by H. K. Clench (1947) and J. Remington (1948), respectively.

It is noteworthy that many important contributors to our knowledge of insects have been amateur entomologists. Of the men discussed here, Herman Strecker was an architect and sculptor, William Henry Edwards was a lawyer and businessman, and Henry Edwards was an actor. The list is long and illustrious, and hopefully men will continue to persevere in this happy tradition.

IN THE FLURRY of naming the *Catocala* during the late nineteenth century, the tendency was to attach a name to every apparently new specimen that came to hand, hoping that time would prove the name worthy of species rank. Accordingly, many forms were originally described as new species: (e.g. "sinuosa," a form of *coccinata*; "gisela," a form of *micronympha*; "aholah," a form of *similis*). In this same vein, there was a tendency to attach new names to very minor variants (e.g., "moderna," a small *maestosa*), and this practice led to the rather embarassing situation of females being described as new forms (e.g., "basalis," the female of *habilis*; "curvata," the female of *robinsoni*; "hinda," the female of *innubens*).

With all this activity, it was inevitable that some species would be described by two or more authors, thereby creating synonyms. The decision as to which name should stand in such a circumstance is based upon the rule of priority (i.e., which name was published first). This rule is ordinarily, and in principle, easy to apply. But given the competitive furor of the times and the personalities involved, considerable controversy did arise, and the ensuing disputes were often heated.

Perhaps the flavor and personalities of that era can best be recalled through excerpts from the writings of some of the principals. For example, the Reverend George D. Hulst in a paper entitled "Remarks Upon the Genus CATOCALA, with a Catalogue of Species and Accompanying Notes" (1880), observed in his introductory comments:

> It is very certain that full knowledge will largely reduce the so-called species. By the necessities of the case, species are at first very largely multiplied. They are very generally based upon a single specimen, often faded, rubbed, or mutilated. . . .
>
> Childhood is troubled with its own diseases. . . . The infancy of a science has a like experience.

This paper was not received with equanimity by A. R. Grote, as the following response makes clear:

> The publication of a paper on the species of *Catocala*, by a clergyman, the Rev. Mr. Geo. D. Hulst . . . obliges me to notice its contents briefly. The criticism that I make on this paper is, that its publication was entirely unnecessary from a scientific point of view. . . .

In a somewhat lengthy preamble, in which I find nothing original which is at the same time important, Mr. Hulst likens the present knowledge of the species of *Catocala* to a diseased infancy. In this Mr. Hulst confounds the state of his own mind on the subject with that of others. . . .

Actually, what really incited Grote's wrath became clear later when Hulst sided with Herman Strecker in one of the interminable conflicts between him and Grote:

We are sorry in our service of science to be compelled to judge between Messrs. Strecker and Grote in a matter which has been so prolific of ill-feeling between them. Both claim priority in the naming of three species of Catocalae. Attempting to get at the truth, irrespective of personal feeling toward either of these gentlemen, to both of whom we are under obligation for favors, we give our judgment in favor of the names of Mr. Strecker. . . . (1880, p. 8)

Grote's views on Strecker are abundantly clear:

Mr. Strecker's work is, on the whole, of such an indifferent character that I am unwilling to criticize it. He has made proportionately more and more unexcusable synonyms than any other writer, and his slovenly descriptions and confessed unacquaintance with structure place him on a level with the worst amateur who has "coined" a "species." In vulgarity and misrepresentation he is, fortunately, without a rival. (1881, p. 157)

Strecker was rarely bested in these matters. He had previously reviewed Grote's "Check List of North American Noctuidae, Part 1" (1875), and concluded his brief statement as follows.

The whole thing is scarcely worth the time devoted to this review, but as the advertisement would lead us to expect quite a different production, than that really furnished, we have given this cursory warning because the price demanded is entirely too big to pay for trunk paper. (1876, Lepid. Rhop. & Het., p. 123)

To return to the situation referred to by Hulst, the question at issue concerned the dates of publication of Grote's descriptions of *C. anna*, *adoptiva*, and *levettei*, and Strecker's descriptions of these same species as *C. amestris*, *delilah*, and *judith*. An accusation of antedating was made:

With regard to the species re-described by Mr. Strecker under the date of "August," whereas the publication was not received until November 12, I have shown that Mr. Strecker placed a false date, and have exposed his motive for doing so. (Grote, 1881, p. 164)

In this particular case, most scholars now agree that Grote had his names first in manuscript, but that Strecker's were published first and so have priority. In fairness, however, it must be noted that questions regarding Strecker's methods were not completely unreasonable (see Barnes & McDunnough, 1918, p. 21).

No doubt the day of lively controversy over the matter of names has passed by — certainly the luxury of such personal and polemical writing is rarely countenanced in today's scientific journals.

IT IS UNFORTUNATE that the book most often consulted for identification of the *Catocala*, W. J. Holland's "Moth Book" (1903), was prepared during this period of controversy and confusion regarding names and synonymy within the genus. Numerous errors were made in that work, and have, unfortunately, been perpetuated in the literature. The recent reprint of the "Moth Book" (1968, Dover, New York) is emended by A. E. Brower, and the *Catocala* appearing in it may now be correctly identified, though Brower occasionally continues Holland's confusing practice of using the names of forms as if they were species (e.g. *Catocala conspicua*, rather than *Catocala ilia*, form "conspicua").

The most elegant work on the North American *Catocala* to appear to date is the memoir by W. Barnes and J. McDunnough, "Illustrations of the North American Species of the Genus *Catocala*" (1918, American Museum of Natural History). This appeared after most of the controversy regarding these moths had been settled, and the watercolor drawings of both larvae and adults (by Mrs. William Beutenmüller and Mr. S. Fred Prince) have not been equalled. Unfortunately, this work is now hard to find, and rather expensive.

In a more technical vein, the best source of information on the *Catocala* is W. T. M. Forbes's "Lepidoptera of New York and Neighboring States. Part III" (1954, Cornell University Agricultural Experiment Station). This work, though not illustrated, provides keys to both larvae and adults, and the species accounts include reference to foodplants and geographic ranges.

In concluding this chapter, a few words might be said regarding the problem that names present to any would-be student of the *Catocala*. How is one to learn the names of so many species and forms? There is, of course, no easy answer.

I would suggest that one's efforts be concentrated on the scientific names, and that these be treated as if they were the common names. Actually, I believe that the scientific names are at least as easy to learn, and that they are, in many cases, less confusing. I have found it particularly helpful to learn something of the derivation of the scientific names, and would suggest some study along these lines as an effective aid to memory. The chief value of the scientific names lies in their universality. For the common names, despite their charm, may vary from time to time and place to place. One man's Sweetheart may be another man's Wife!

My name is Legion: for we are many.

Mark 5: 9

A Fertile Theme

II

THE CATOCALA comprise an exceedingly complex assemblage of closely related moths. The simple theme of somber forewings and striking hindwings has yielded a seemingly infinite array of variations. One's initial response to this profusion is astonishment. But if one's interest is aroused, astonishment turns quickly to despair at the matter of names. Species after species, form after form, and subspecies, and aberrations . . . and each one has a name!

The magnitude of this problem can be appreciated when one realizes that there are at least 200 *Catocala* species worldwide. The genus is predominantly North Temperate (Holarctic), and its metropolis is clearly North America (Nearctic). McDunnough (1938) lists 104 North American species (and 136 additional named varieties), while some 80 to 90 species are known from Europe and Asia (Palearctic) (Seitz, 1914). Even in the restricted area covered in the present book (North America east of the Mississippi River), there are at least 71 *Catocala* species and nearly 100 more named varieties.

Such a complex assemblage poses substantial difficulties in terms of the identification of its members. A primary aim of the present chapter is to provide a basis for the identification of most of the eastern *Catocala*. This will be accomplished by means of color-plates of specimens and a series of species accounts.

Since the color-plates will provide the initial guide to identification, some remarks concerning them are appropriate. First, most *Catocala* exhibit considerable variation and no one specimen will be entirely representative of its species. The sex of a specimen is a factor that must be considered, for females are often more boldly and heavily marked than males (fig. 2.1D). Specimens may also change with age. In particular, older specimens may fade or take on a brownish cast that is not characteristic of fresh specimens. Such sources of variation and many others (fig. 2.1), should dictate some caution in using the plates as a sole basis for identification. They are best used in conjunction with the species accounts, where critical details regarding similar species, ranges, seasons, and status may be found.

A
B
C

Figure 2.1 Examples of variation within various species: (A) *C. ilia*, contrast differences (*left* — uniform; *right* — highly patterned); (B) *C. lacrymosa*, polymorphism (*left* — form "zelica"; *right* — form "paulina"); (C) *C. relicta* "clara," prominence of lines (*left* — lines obsolete; *right* — lines prominent); (D) *C. habilis*, sex differences (*left* — male, no prominent dashes; *right* — female, prominent basal and anal dashes); (E) *C. retecta*, size differences (*left* — small male; *right* — large female). 0.8X.

Plate 1

1 *C. innubens*, ♂. Hunterdon Co., New Jersey; ex larva, 1973; J. Muller (DFS). p. 37.
2 *C. innubens* "scintillans," ♀. Hunterdon Co., New Jersey; ex larva, 1974; J. Muller (TDS). p. 37.
3 *C. innubens*, ♀. Morenci, Lenawee Co., Michigan; 23 August 1974; M. C. Nielsen (TDS). p. 37.
4 *C. muliercula*, ♂. Delmont, Cumberland Co., New Jersey; August 1974; C. Brooke Worth & J. Muller (TDS). p. 39.
5 *C. coelebs*, ♂. Lake Kejimukujik, Queens Co., Nova Scotia; 6 August 1961; D. C. Ferguson (USNM). p. 40.
6 *C. badia*, ♀. Harwichport, Barnstable Co., Massachusetts; 7 October 1972; T. D. Sargent (TDS). p. 40.
7 *C. antinympha*, ♂. Leverett, Franklin Co., Massachusetts; ex larva, 1968; T. D. Sargent (TDS). p. 39.
8 *C. habilis*, ♂. Leverett, Franklin Co., Massachusetts; 20 August 1970; T. D. Sargent (TDS). p. 41.
9 *C. judith*, ♂. Leverett, Franklin Co., Massachusetts; 26 July 1971; T. D. Sargent (TDS). p. 44.
10 *C. serena*, ♂. Resica Falls, Monroe Co., Pennsylvania; 25 August 1971; D. F. Schweitzer (DFS). p. 41.
11 *C. epione*, ♀. Leverett, Franklin Co., Massachusetts; 25 July 1971; T. D. Sargent (TDS). p. 38.
12 *C. miranda*, ♀. Fontana Dam, Graham Co., North Carolina; 9 July 1972; D. F. Schweitzer (DFS). p. 65.
13 *C. consors* "sorsconi," ♀. Missouri; 23 July 1918; (PMNH). p. 38.
14 *C. piatrix*, ♀. Strafford, Chester Co., Pennsylvania; 23 September 1970; D. F. Schweitzer (DFS). p. 37.
15 *C. consors*, ♀. Archbold Biological Station, Highlands Co., Florida; 22 May 1958; R. W. Pease, Jr. (PMNH). p. 38.

0.80X

Plate 2

1 *C. obscura,* ♂. Leverett, Franklin Co., Massachusetts; 25 August 1971; T. D. Sargent (TDS). p. 46.
2 *C. residua,* ♂. Leverett, Franklin Co., Massachusetts; 15 August 1972; T. D. Sargent (TDS). p. 46.
3 *C. robinsoni,* ♂. Mansfield, Tolland Co., Connecticut; 6 September 1958; H. P. Wilhelm (PMNH). p. 44.
4 *C. robinsoni* "missouriensis," ♂. Sleepy Creek, Edgefield Co., South Carolina; August 1954; J. Bauer (CM). p. 44.
5 *C. angusi,* ♂. Dupo, St. Clair Co., Illinois; 16 August 1928; F. R. Arnhold (PMNH). p. 45.
6 *C. angusi* "lucetta," ♂. Sleepy Creek, Edgefield Co., South Carolina; August 1954; J. Bauer (CM). p. 45.
7 *C. insolabilis,* ♀. J. Doll (JBP). p. 49.
8 *C. flebilis,* ♂. Leverett, Franklin Co., Massachusetts; 19 August 1972; T. D. Sargent (TDS). p. 45.
9 *C. agrippina,* ♂. Rye, Orange Co., Florida; 22 April 1919; W. Reiff (JBP). p. 47.
10 *C. vidua,* ♀. Morenci, Lenawee Co., Michigan; 23 August 1959; M. C. Nielsen (MCN). p. 50.
11 *C. sappho,* ♂. Rye, Orange Co., Florida; 9 May 1919; W. Reiff (PMNH). p. 47.
12 *C. maestosa,* ♀. 11 June 1917; (PMNH). p. 50.

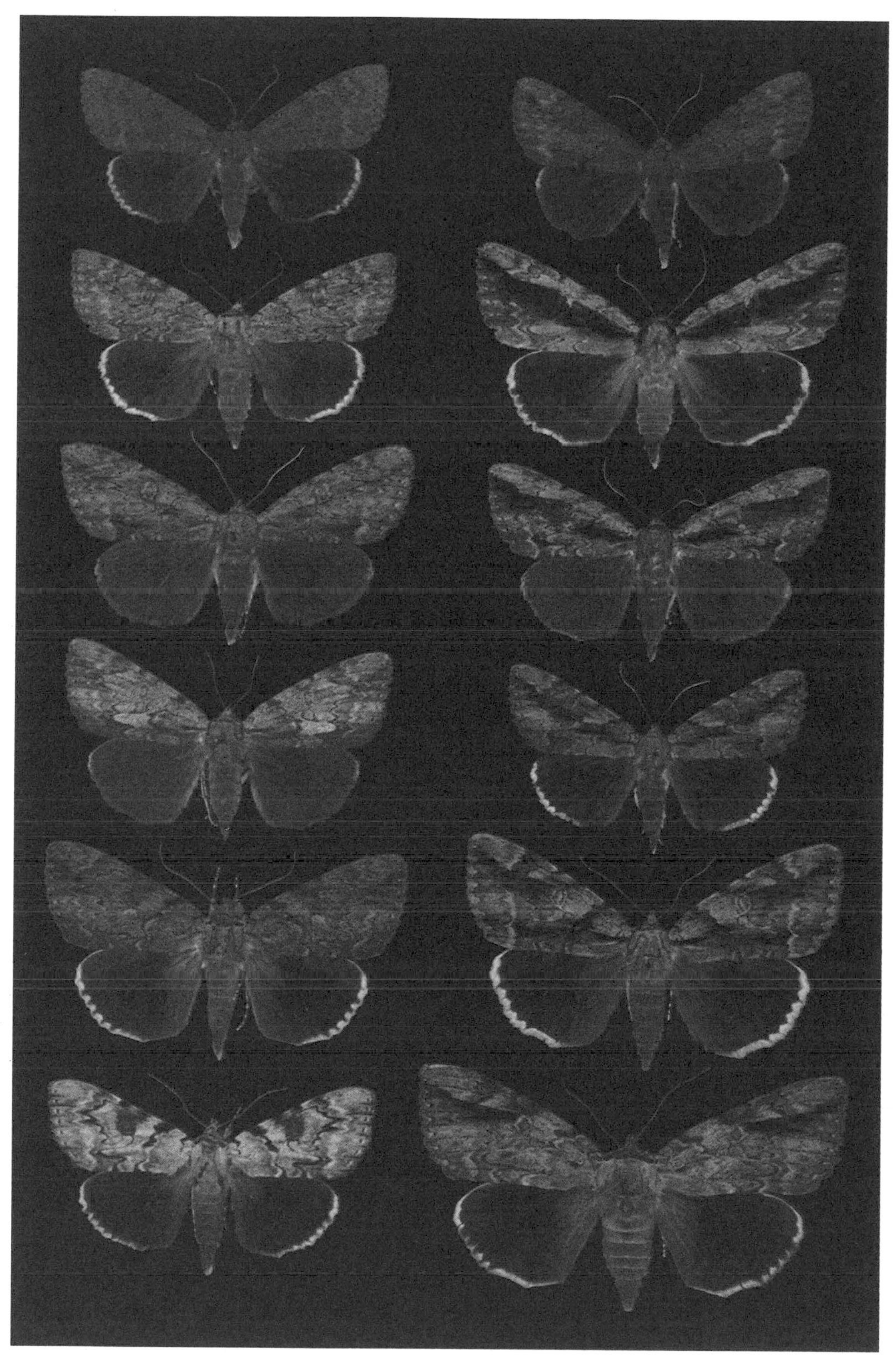

0.80X

Plate 3

1 *C. retecta,* ♂. Leverett, Franklin Co., Massachusetts; 27 August 1974; T. D. Sargent (TDS). p. 48.
2 *C. retecta* "luctuosa," ♀. Sleepy Creek, Edgefield Co., South Carolina; August 1954; J. Bauer (CM). p. 48.
3 *C. dejecta,* ♂. Leverett, Franklin Co., Massachusetts; 2 August 1973; T. D. Sargent (TDS). p. 49.
4 *C. ulalume,* ♂. (PMNH). p. 48.
5 *C. lacrymosa* "zelica," ♂. Eureka, St. Louis Co., Missouri; 7 August 1937; F. R. Arnhold (PMNH). p. 51.
6 *C. lacrymosa,* ♂. Missouri; 28 August 1932; (TDS). p. 51.
7 *C. lacrymosa* "paulina," ♂. Ranken, St. Louis Co., Missouri; P. S. & C. L. Remington (PMNH). p. 51.
8 *C. lacrymosa* x *palaeogama* (?), ♂. Morenci, Lenawee Co., Michigan; 1 September 1973; M. C. Nielsen (MCN). p. 51.
9 *C. palaeogama* "annida," ♀. Washington, Litchfield Co., Connecticut; 31 July 1971; S. A. Hessel (TDS). p. 51.
10 *C. palaeogama,* ♂. Leverett, Franklin Co., Massachusetts; 14 August 1972; T. D. Sargent (TDS). p. 51.
11 *C. palaeogama* "phalanga," ♀. Washington, Litchfield Co., Connecticut; 31 July 1962; S. A. Hessel (TDS). p. 51.
12 *C. palaeogama* "denussa," ♂. Washington, Litchfield Co., Connecticut; 10 August 1962; S. A. Hessel (TDS). p. 51.

0.80X

Plate 4

1 *C. nebulosa,* ♂. St. Joseph Co., Michigan; 14 August 1971; M. C. Nielsen (MCN). p. 52.
2 *C. subnata,* ♂. Washington, Litchfield Co., Connecticut; 8 August 1967; S. A. Hessel (TDS). p. 52.
3 *C. neogama,* ♀. Leverett, Franklin Co., Massachusetts; 8 August 1967; T. D. Sargent (TDS). p. 53.
4 *C. neogama,* ♂. Leverett, Franklin Co., Massachusetts; 19 July 1971; T. D. Sargent (TDS). p. 53.
5 *C. ilia,* ♀. Leverett, Franklin Co., Massachusetts; 15 July 1971; T. D. Sargent (TDS). p. 53.
6 *C. ilia* "conspicua," ♀. Leverett, Franklin Co., Massachusetts; 21 July 1971; T. D. Sargent (TDS). p. 53.
7 *C. ilia* "satanas," ♀. Leverett, Franklin Co., Massachusetts; 23 July 1971; T. D. Sargent (TDS). p. 53.
8 *C. ilia* "normani," ♀. Leverett, Franklin Co., Massachusetts; 17 July 1968; T. D. Sargent (TDS). p. 53.
9 *C. relicta* "clara," ♀. Leverett, Franklin Co., Massachusetts; 30 July 1971; T. D. Sargent (TDS). p. 54.
10 *C. relicta,* ♀. Sunderland, Franklin Co., Massachusetts; 28 July 1965; T. D. Sargent (TDS). p. 54.
11 *C. relicta* "phrynia," ♂. Leverett, Franklin Co., Massachusetts; 30 July 1971; T. D. Sargent (TDS). p. 54.
12 *C. relicta* "clara," ♂. Leverett, Franklin Co., Massachusetts; ex ova, experimentally painted hindwings; 1970; (TDS). p. 54.

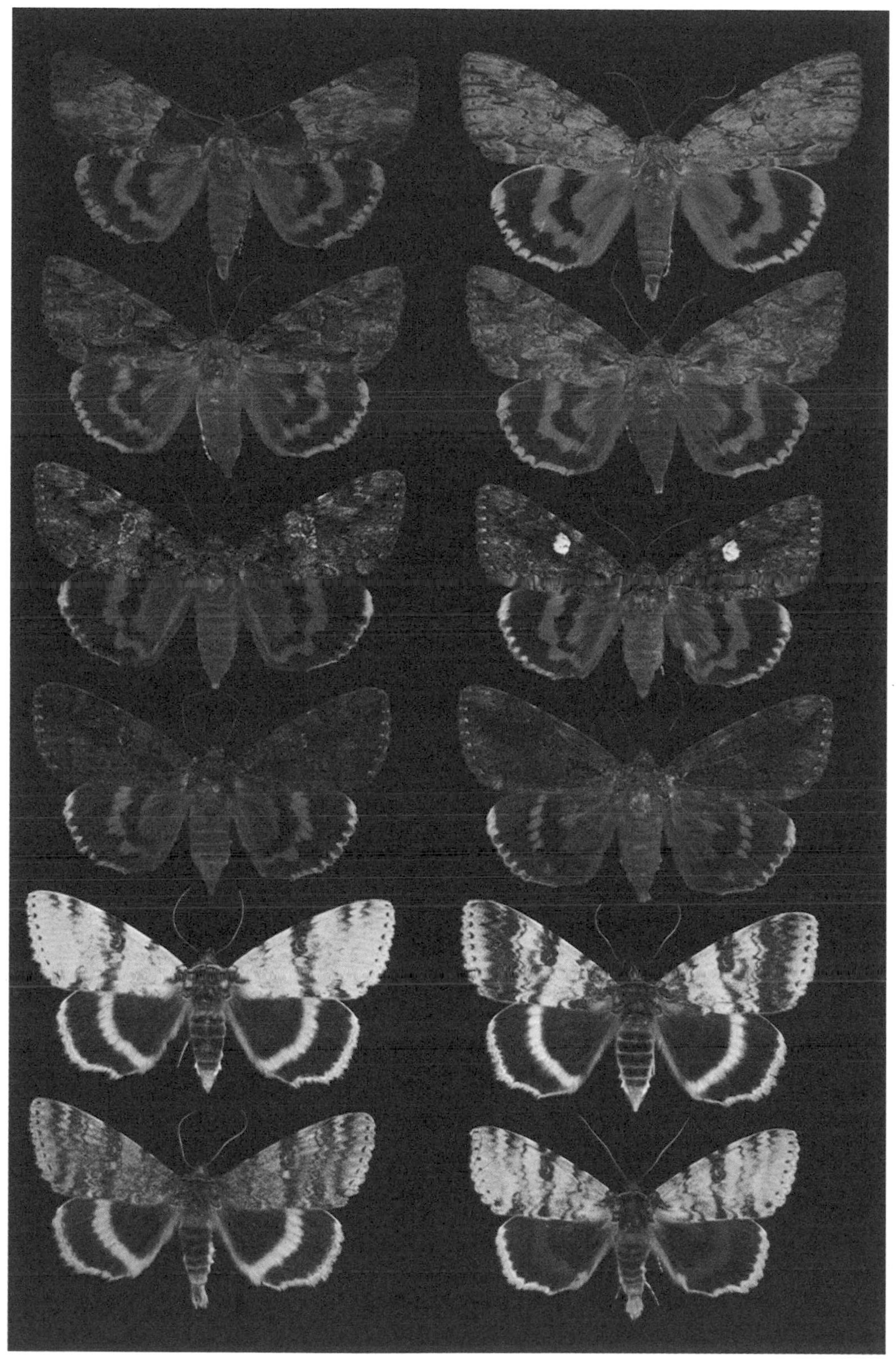

0.80X

Plate 5

1 *C. cerogama,* ♀. Leverett, Franklin Co., Massachusetts; 16 August 1971; T. D. Sargent (TDS). p. 54.
2 *C. cerogama* "ruperti," ♀. Paratype. Ithaca, Tompkins Co., New York; 20 August 1937; J. G. Franclemont (USNM). p. 54.
3 *C. marmorata,* ♂. New York; (PMNH). p. 55.
4 *C. meskei,* ♂. Columbus, Platte Co., Nebraska; 9 July 1958; E. A. Froemel (PMNH). p. 57.
5 *C. unijuga,* ♀. Leverett, Franklin Co., Massachusetts; 22 July 1968; T. D. Sargent (TDS). p. 55.
6 *C. unijuga* "agatha," ♂. Camden, Knox Co., Maine; 27 August 1970; D. F. Schweitzer (DFS). p. 55.
7 *C. briseis,* ♂. Meeme, Manitowoc Co., Wisconsin; 14 August 1972; C. G. Kellogg (CGK). p. 56.
8 *C. briseis,* ♂. Sunderland, Franklin Co., Massachusetts; 30 July 1966; T. D. Sargent (TDS). p. 56.
9 *C. semirelicta* "atala," ♂. Otsego Co., Michigan; 13 August 1959; M. C. Nielsen (MCN). p. 57.
10 *C. semirelicta,* ♂. Clare Co., Michigan; 22 August 1956; M. C. Nielsen (MCN). p. 57.
11 *C. parta,* ♂. Halifax, Nova Scotia; 16 August 1959; D. C. Ferguson (USNM). p. 56.
12 *C. parta* "forbesi," ♀. Finleyville, Washington Co., Pennsylvania; 1–7 July; (CM). p. 56.

0.80X

Plate 6

1 C. *junctura,* ♂. Creve Coeur L., St. Louis Co., Missouri; 14 July 1934; P. S. Remington (PMNH). p. 58.
2 C. *cara,* ♂. Leverett, Franklin Co., Massachusetts; 14 August 1969; T. D. Sargent (TDS). p. 58.
3 C. *amatrix* "hesseli," ♂. Woodmere, Nassau Co., New York; 21 August 1932; S. A. Hessel (PMNH). p. 59.
4 C. *amatrix* x *cara* (?), ♂. West Hatfield, Hampshire Co., Massachusetts; 23 August 1970; C. G. Kellogg (CGK). pp. 58–59.
5 C. *concumbens,* ♂. Leverett, Franklin Co., Massachusetts; 26 August 1971; T. D. Sargent (TDS). p. 59.
6 C. *amatrix* "selecta," ♂. West Hatfield, Hampshire Co., Massachusetts; 24 August 1970; C. G. Kellogg (CGK). p. 59.
7 C. *coccinata,* ♂. Leverett, Franklin Co., Massachusetts; 16 July 1971; T. D. Sargent (TDS). p. 65.
8 C. *amatrix,* ♂. West Hatfield, Hampshire Co., Massachusetts; 2 September 1970; C. G. Kellogg (CGK). p. 59.
9 C. *ultronia,* ♀. Leverett, Franklin Co., Massachusetts; 26 July 1971; T. D. Sargent (TDS). p. 66.
10 C. *herodias gerhardi,* ♀. Lakehurst, Ocean Co., New Jersey; ex ova, 11 July 1951; D. C. Ferguson (USNM). p. 64.
11 C. *ultronia* "celia," ♂. Leverett, Franklin Co., Massachusetts; 27 July 1974; T. D. Sargent (TDS). p. 66.
12 C. *ultronia* "lucinda," ♂. Pelham, Hampshire Co., Massachusetts; 24 July 1965; T. D. Sargent (TDS). p. 66.
13 C. *ultronia* "nigrescens," ♂. Leverett, Franklin Co., Massachusetts; 28 July 1971; T. D. Sargent (TDS). p. 66.

0.80X

Plate 7

1 *C. delilah,* ♂. Terrell, Kaufman Co., Texas; 8 June 1941; V. G. Sasko (PMNH). p. 60.
2 *C. illecta,* ♂. Ottawa, Franklin Co., Kansas; 13 June 1950; W. H. Howe (PMNH). p. 60.
3 *C. abbreviatella,* ♀. Hill City, Pennington Co., South Dakota; 19 July 1964; D. C. Ferguson (PMNH). p. 61.
4 *C. amestris,* ♀. McClellanville, Charleston Co., South Carolina; 20 June 1971; R. B. Dominick & C. R. Edwards (USNM). p. 61.
5 *C. clintoni,* ♀. Missouri; 9 June 1902; Barnes collection (USNM). p. 71.
6 *C. nuptialis,* ♀. Dresbach, Winona Co., Minnesota; 27 July 1938; F. R. Arnhold (PMNH). p. 61.
7 *C. whitneyi,* ♂. Columbus, Platte Co., Nebraska; 5 July 1958; E. A. Froemel (PMNH). p. 61.
8 *C. messalina,* ♀. Crescent Beach, St. Johns Co., Florida; 29 May 1954; J. Bauer (CM). p. 62.
9 *C. titania,* ♂. Madison, St. Louis Co., Illinois; 27 June 1937; F. R. Arnhold (PMNH). p. 70.
10 *C. alabamae,* ♂. Springfield, Texas; (PMNH). p. 70.
11 *C. dulciola,* ♂. Barnes collection (USNM). p. 71.
12 *C. sordida,* ♀. Leverett, Franklin Co., Massachusetts; 6 August 1970; T. D. Sargent (TDS). p. 62.
13 *C. sordida,* ♂. Big Indian L., Halifax Co., Nova Scotia; 2 August 1958; D. C. Ferguson (PMNH). p. 62.
14 *C. gracilis* "cinerea," ♂. Pelham, Hampshire Co., Massachusetts; 26 July 1964; T. D. Sargent (TDS). p. 63.
15 *C. sordida,* ♂. Leverett, Franklin Co., Massachusetts; 27 July 1974; T. D. Sargent (TDS). p. 62.
16 *C. sordida* "metalomus," ♂. Boulderwood, Halifax Co., Nova Scotia; 12 August 1959; D. C. Ferguson (PMNH). p. 62.
17 *C. gracilis,* ♀. Leverett, Franklin Co., Massachusetts; 22 July 1973; T. D. Sargent (TDS). p. 63.
18 *C. louiseae,* ♀. Paratype. Moultrie, St. Johns Co., Florida; 14 May 1954; J. Bauer (CM). p. 63.
19 *C. gracilis* "herperis," ♂. Woodmere, Nassau Co., New York; 10 July 1951; S. A. Hessel (PMNH). p. 63.
20 *C. gracilis* "lemmeri," ♀. Lakehurst, Ocean Co., New Jersey; 1 August; F. Lemmer (USNM). p. 63.
21 *C. andromedae,* ♂. Leverett, Franklin Co., Massachusetts; 28 July 1972; T. D. Sargent (TDS). p. 64.
22 *C. grynea,* ♂. Leverett, Franklin Co., Massachusetts; 25 July 1971; T. D. Sargent (TDS). p. 69.
23 *C. praeclara,* ♂. Pelham, Hampshire Co., Massachusetts; 17 July 1964; T. D. Sargent (TDS). p. 69.

0.80X

Plate 8

1 *C. crataegi,* ♀. Pelham, Hampshire Co., Massachusetts; 10 July 1965; T. D. Sargent (TDS). pp. 66–67.
2 *C. mira,* ♂. Leverett, Franklin Co., Massachusetts; 28 July 1971; T. D. Sargent (TDS). p. 68.
3 *C. blandula,* ♀. Leverett, Franklin Co., Massachusetts; July 1971; T. D. Sargent (TDS). p. 68.
4 *C. "pretiosa,"* ♀. Manchester, Hillsborough Co., New Hampshire; (JBP). pp. 66–67.
5 *C. connubialis,* ♀. Lake Kejimukujik, Queens Co., Nova Scotia; 6 August 1961; D. C. Ferguson (USNM). p. 73.
6 *C. blandula,* ♀. Leverett, Franklin Co., Massachusetts; 11 July 1969; T. D. Sargent (TDS). p. 68.
7 *C. minuta,* ♂. Strafford, Chester Co., Pennsylvania; 8 August 1970; D. F. Schweitzer (DFS). p. 72.
8 *C. connubialis* "cordelia," ♂. Fontana Dam, Graham Co., North Carolina; 9 July 1971; D. F. Schweitzer (DFS). p. 73.
9 *C. micronympha,* ♀. Washington, Litchfield Co., Connecticut; 13 July 1971; S. A. Hessel (PMNH). p. 74.
10 *C. minuta* "parvula," ♂. Strafford, Chester Co., Pennsylvania; 28 June 1970; D. F. Schweitzer (DFS). p. 72.
11 *C. connubialis* "cordelia-pulverulenta," ♀. Leverett, Franklin Co., Massachusetts; 12 July 1973; T. D. Sargent (TDS). p. 73.
12 *C. micronympha* "hero," ♀. Ressica Falls, Monroe Co., Pennsylvania; 20 July 1971; D. F. Schweitzer (DFS). p. 74.
13 *C. minuta* "mellitula," ♂. New York; 13 July 1918; (TDS). p. 72.
14 *C. connubialis* "pulverulenta," ♀. West Hatfield, Hampshire Co., Massachusetts; 29 July 1972; C. G. Kellogg (CGK). p. 73.
15 *C. micronympha* "hero," ♀. Leverett, Franklin Co., Massachusetts; 1 August 1970; T. D. Sargent (TDS). p. 74.
16 *C. minuta* "obliterata," ♀. Valley Forge, Pennsylvania; ex larva, 6 June 1973; D. F. Schweitzer (DFS). p. 72.
17 *C. connubialis* "broweri," ♂. Leverett, Franklin Co., Massachusetts; 19 July 1971; T. D. Sargent (TDS). p. 73.

18 *C. micronympha* "gisela," ♀. Leverett, Franklin Co., Massachusetts; 14 July 1971; T. D. Sargent (TDS). p. 74.
19 *C. grisatra,* ♀. Allotype. Athens, Clarke Co., Georgia; 24 June 1926; B. Maguire (CM). p. 73.
20 *C. similis,* ♀. Pelham, Hampshire Co., Massachusetts; 24 July 1965; T. D. Sargent (TDS). p. 72.
21 *C. micronympha* "lolita," ♀. Washington, Litchfield Co., Connecticut; 6 July 1962; S. A. Hessell (PMNH). p. 74.
22 *C. amica,* ♀. Leverett, Franklin Co., Massachusetts; 16 August 1970; T. D. Sargent (TDS). p. 75.
23 *C. amica,* ♂. Leverett, Franklin Co., Massachusetts; 26 July 1970; T. D. Sargent (TDS). p. 75.
24 *C. amica* "suffusa," ♂. Orlando, Orange Co., Florida; 20 April [before 1940]; (TDS). p. 75.
25 *C. amica* "curvifascia," ♂. Washington, Litchfield Co., Connecticut; 21 July 1969; S. A. Hessel (TDS). p. 75.
26 *C. jair,* ♀. Flagler Co., Florida; 29 May 1954; J. Bauer (CM). p. 75.
27 *C. jair* (?), ♀. Batsto, Burlington Co., New Jersey; 22 July 1972; D. F. Schweitzer (DFS). p. 75.

0.80X

The species accounts, in addition to assisting in identification, provide considerable information about each species in a reasonably orderly fashion. Each includes the following sections: description, similar species, range, status, season, larval foodplants, and notes. A few remarks about the material in each of these sections follows.

Description. The moth and its most conspicuous varieties are briefly described, with emphasis on its consistent and distinctive features. The description includes terms and abbreviations which refer to various parts and markings of the wings, as depicted in figure 2.2. The size (distance in mm. across the spread forewings) is included, as are references to color illustrations of the species in Holland's *Moth Book* (**H**) and Barnes & McDunnough's *Illustrations of the North American Species of the Genus Catocala* (**B** & **Mc**).

Similar species. This section lists other species or varieties with which the species under consideration is most easily confused, and indicates the most reliable distinguishing characteristics.

Range. The geographic range over which the species is known to occur is given in terms of the north and south limits of distribution at the eastern edge of the range, followed by the western limit (if east of the Mississippi River). It should be noted that *Catocala* ranges are difficult to specify, as individuals may wander long distances, and changing climatic conditions may be accompanied by range extensions and contractions over the years.

Status. The present abundance of the species over its range is given, insofar as it can be determined from collectors and collections, and always bearing in mind that the size of *Catocala* populations may vary widely from year to year.

Season. An indication of the usual time of appearance of the adults is given, usually expressed as "early," "mid-season," or "late," with reference to the entire *Catocala* season. For species taken in numbers in southern New England (based on records of Hessel and Sargent, 1961–1973), the median and extreme dates of capture are given. These dates may be used as indicators of the relative seasonal occurrence of various species.

Larval foodplants. The plants upon which the larvae are *known* to feed are indicated, i.e., plants on which successful development to the adult stage has been reported. Plants upon which larvae have been found, but not reared, are not included. No life history data are reported, since these can be found in literature referred to by H. M. Tietz (1972).

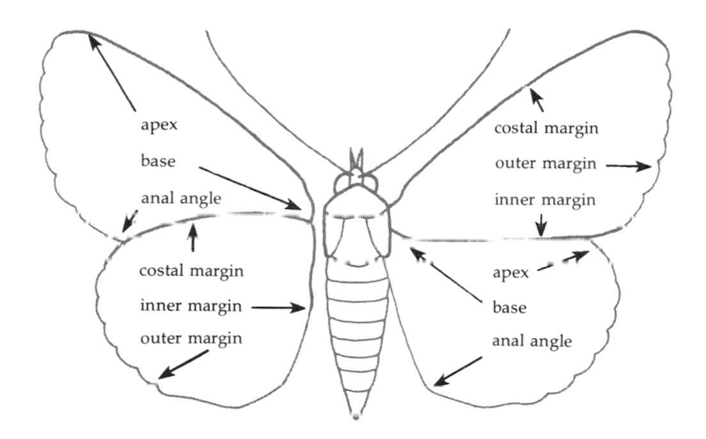

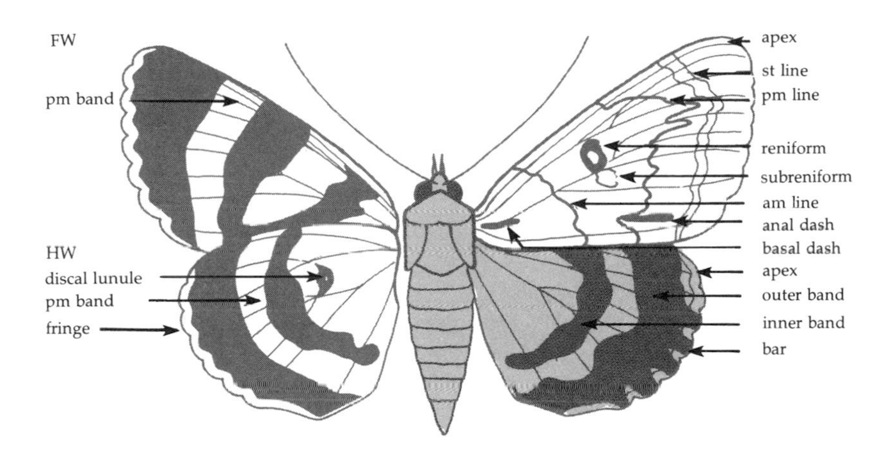

Figure 2.2 Various parts (A) and markings (B) of *Catocala* wings. Abbreviations: FW (forewing), HW (hindwing), am (antemedial), pm (postmedial), st (subterminal).

Notes. Studies and observations on the species, particularly those relating to adult behavior, are briefly noted.

The species are considered according to their arrangement in the latest check-list of North American moths (McDunnough, 1938). The genus will undoubtedly undergo a taxonomic revision in due course, and this almost certainly will result in considerable re-arrangement of the species. But it seems most prudent to use a familiar arrangement at the present time, leaving any changes to competent systematists with a comprehensive overview of the genus.

Throughout these species accounts I have drawn heavily on the observations of others. My debt to these individuals is great, and I have attempted to give credit to them (initials in parentheses) at appropriate points in the text. Among the individuals so credited are: J. Bauer, A. E. Brower, R. B. Dominick, D. C. Ferguson, J. G. Franclemont, S. A. Hessel, R. R. Keiper, C. G. Kellogg, W. A. Miller, J. Muller, M. C. Nielson, C. L. Remington, and D. F. Schweitzer.

Figure 2.3 The frenulum (coupling structure holding FW and HW together) as it appears in males (*left*) and females (*right*). In the male, this structure is a single spine, engaging near the costa; in the female, it is multiple, engaging nearer the center of the forewing. This difference is the most reliable indicator of sex in the *Catocala*.

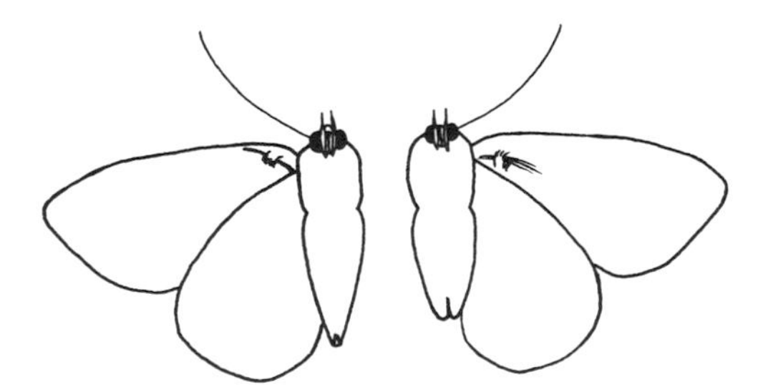

C. innubens Guenée (1852)

THE BETROTHED

Plate 1: 1, 2, 3

Description. FW in male rather uniform gray-brown; first quarter of pm line often whitish, forming shallow arc from costa; often with heavy black streak between subreniform and pm line. Underside FW with cream, not orange, pm band. HW deep orange with black bands. Underside HW with orange ground on lower half, cream on upper half. Female, with dark shadings on FW from am line through subreniform and reniform area, sometimes extending along costa, is "hinda" French. Form with very dark, contrasting area between am and pm lines of FW is "scintillans" Grote. Aberration with yellow HW (perhaps produced by rearing under hot, dry conditions) is "flavidalis" Grote. 55–65 mm. **H** 33: 9 ("scintillans"), 10 (female as "hinda"), 13. **B** & **Mc** 7: 9, 10, 11 ("scintillans").

Similar species. Other species having rich brown FW cast (e.g., *muliercula, badia, piatrix*) usually have *yellow*-orange HW ground. May be occasionally confused with some *ilia*, which usually have much redder HW ground.

Range. Southern Canada (Ontario) to Florida and west throughout area, where foodplant occurs. Rare to the NE in this range, absent from Maine (AEB).

Status. Uncommon to common in vicinity of foodplant, rare elsewhere.

Season. Mid-season to late. (Southern New England: 22 August; 31 July–20 September.)

Larval foodplants. Honey Locust (*Gleditsia triacanthos*).

Notes. Moth rests head-down on tree trunks, usually from 3 to 9 ft. above ground, and quite often under leaves of Poison Ivy or Virginia Creeper (MCN).

C. piatrix Grote (1864)

THE PENITENT

Plate 1: 14

Description. FW dark, rather uniform brown-gray; some lightening from am line to near reniform along costa, and extending inward to include the usually open subreniform; am line with characteristic "dip" above inner margin. HW orange (not reddish) with black bands and dark basal hairs. Underside largely fuscous and pale ochre-orange, with black median band on both FW and HW. 70–80 mm. **H** 36: 6. **B** & **Mc** 6: 2, 3 (pale, western specimen is "dionyza" H. Edwards).

Similar species. See *innubens*. Some dark *neogama* may be similar, but pm orange band on HW much narrower, and subreniform of FW usually closed, in that case. *C. ilia* "confusa" has a reddish HW ground.

Range. Southern Canada (Ontario), Maine (rare) to Florida and west throughout area.

Status. Uncommon to common throughout most of range; more common in and near cities (where hickories and walnuts planted) to the north.

Season. Late. (Southern New England: 1 September; 20 August–9 September.)

Larval foodplants. Walnuts (*Juglans*) and hickories (*Carya*), apparently preferring Black Walnut (*J. nigra*).

Notes. Full-grown larvae may be found resting in grass at base of foodplant (SAH).

C. *consors* (Smith & Abbot) (1797)
THE CONSORT
Plate 1: 13, 15

Description. FW dull gray, with slight brownish shading along the single, prominent am and pm lines; some darkening between subreniform-reniform area and pm line. HW yellow-orange with black bands; pm orange band narrow and very irregular ("zigzag"). The name "sorsconi" Barnes & Benjamin is usually applied to northern specimens (i.e., north of Gulf States) which have more black, and a more zigzag pm band on HW. 60–70 mm. **H** 34: 3. **B & Mc** 7: 7.

Similar species. HW pattern will distinguish from any vaguely similar species (e.g., *coelebs*, *delilah*).

Range. Long Island, New York to Florida and west throughout area. (One of the types of "sorsconi" was labelled Maine.)

Status. Generally rare to NE; uncommon to common elsewhere.

Season. Early to mid-season.

Larval foodplants. Hickories (*Carya*) (AEB).

Notes. Larvae seem to prefer small trees, and may rest on foodplant when full-grown (AEB, JGF, SAH).

C. *epione* (Drury) (1770)
EPIONE UNDERWING
Plate 1: 11

Description. FW blackish-gray, with prominent brown shading along am and pm lines; st whitish and contrasting. HW solid black with pure white fringe (no barring). Underside HW blackish, with only trace of pm white band toward costa (fig. 2.4C). Some near-melanic individuals occur, with no trace of whitish on FW. 55–65 mm. **H** 31: 3. **B & Mc** 1: 16.

Similar species. The pure white HW fringe is quite distinctive.

Range. Maine and southern Canada (Ontario) to Florida and west throughout area.

Status. Generally common to abundant.

Season. Early to mid-season. (Southern New England: 1 August; 9 July–26 September.)

Larval foodplants. Hickories (*Carya*). Reared on Shagbark Hickory (*C. ovata*) (TDS).

Notes. Moth very active; often found resting low on stumps, trunks, and in woodpiles. Also rests under eaves of buildings. When disturbed, may fly to ground. Comes readily to baits.

C. muliercula Guenée (1852)
THE LITTLE WIFE
Plate 1: 4

Description. FW deep, rich brown; quite uniform, with slight darkening within am line and in patches along outer margin (particularly below apex). HW yellow-orange with black bands; outer band very broad, and inner band set in considerably; fringe fuscous except at apex. The aberration "peramans" Hulst has HW almost entirely black. 60–70 mm. **H** 32: 11. **B** & **Mc** 7: 24.

Similar species. See *innubens*. The arrangement of bands on HW will distinguish from anything similar.

Range. Connecticut to Florida, and thence west to Texas.

Status. Uncommon to common in most places with foodplant.

Season. Early to rather late (long season).

Larval foodplant. Wax Myrtle (Bayberry) (*Myrica cerifera*).

Notes. Rests as full-grown larva on foodplant (SAH).

C. antinympha (Hübner) (1852)
WAYWARD NYMPH
SWEET-FERN UNDERWING
Plate 1: 7

Description. FW black, with velvety black lines, and brownish shadings (especially outside pm line). HW orange-yellow with black bands; fringe infuscated as in *muliercula*. Underside light orange and fuscous-black on both FW and HW. A minor variant, with whitish subreniform, is "multoconspicua" Reiff. 45–55 mm. **H** 32: 8. **B** & **Mc** 7: 15.

Similar species. Black FW distinguishes from all but extreme melanics of other species, and these usually show considerable extension of black on HW.

Range. Southern Canada (Nova Scotia) to mid-Atlantic states; chiefly along coast, but extending west through area with foodplant.

Status. Common to abundant along coast, becoming uncommon to rare and sporadic elsewhere.

Season. Early to mid-season, with a few late individuals (sometimes these surprisingly fresh; perhaps a partial second brood). (Southern New England: 1 August; 11 July–26 September.)

Larval foodplant. Sweet-fern (*Comptonia peregrina*).

Notes. Rests as full-grown larva on foodplant. Moths may be flushed from sweet-fern patches (perhaps females), but also found resting head-down on tree trunks; prefer dark backgrounds in experimental tests (Sargent 1966, 1968, 1969 *a*).

C. coelebs Grote (1874)
THE OLD-MAID
Plate 1: 5

Description. FW gray, with brown within am line and along pm line. HW orange-yellow with rather even black bands. 50-60 mm. **H** 32: 18. **B & Mc** 7: 8.

Similar species. FW somewhat similar to *consors*, but pm orange band of HW narrow and zigzag in that species. The gray ground between am and pm lines of FW should distinguish from *badia* (but see "phoebe").

Range. Northern; southern Canada (Nova Scotia, Quebec, Ontario), Maine, New Hampshire, and west to Wisconsin; southward to Adirondacks in New York.

Status. Very local; generally rare.

Season. Early to mid-season.

Larval foodplant. Sweet Gale (*Myrica gale*).

C. badia Grote & Robinson (1866)
BAY UNDERWING BADIA UNDERWING
Plate 1: 6

Description. FW buff-brown, with deeper brown in basal area and within pm line (sometimes nearly filling area between am and pm lines); contrastingly paler beyond pm line; lines faint, barely traceable. HW yellow-orange with black bands; often with terminal yellow-orange line at base of fringe. Specimens from New Hampshire may appear transitional to *coelebs*, having the sharp bend of upper part of pm area and the more distinct lines of that species; this is "phoebe" H. Edwards. 50-60 mm. **H** 32: 10. **B & Mc** 7: 16.

Similar species. See *muliercula* and *coelebs*. Neither of these species has the pm boundary area cutting straight across the FW.

Range. Largely coastal, from Portland, Maine, to New Jersey.

Status. May be very common to abundant along coast, where foodplant abounds; less and less common inland.

Season. Mid-season. (Southern New England: 1 August; 19 July-4 September.) (A late specimen taken on 7 October 1972 at Harwichport, Mass., was very fresh, raising the question, as with *antinympha*, of a partial second brood.)

Larval foodplant. Wax Myrtle (Barberry) (*Myrica cerifera*).

Notes. May be flushed from foodplant patches, and rests on ground in at least some instances.

C. *habilis* Grote (1872)
Plate 1: 8

Description. FW light gray, with slightly contrasting thin black lines, and whitish shadings. Usually with fairly prominent anal dash. Female, more mottled, and with prominent basal dash, is "basalis" Grote. (Some males may also show this basal dash.) HW orange with black bands, and dark brown basal hairs. Underside light orange, with mostly blackish, strongly contrasting bands. A melanic specimen, "denussa" Ehrman, may be of this species, though most authorities feel it is a melanic of *palaeogama*. However, melanics of *habilis* have been reared by AEB. An aberration with mostly black HW is described as "depressans" in chapter five. 55–65 mm. **H** 33: 11, 12 (female as "basalis"). **B** & **Mc** 7: 5.

Similar species. Closest to *serena*, but that species with "smoother" FW, less dentate lines; also prominent, contrastingly brownish collar on thorax; and much more blackish on underside.

Range. Southern Canada (Quebec, Ontario), Maine (rare) to North Carolina, and west throughout area.

Status. Generally uncommon to common, but occasionally abundant.

Season. Late. (Southern New England: 12 September; 11 August–21 October.)

Larval foodplants. Hickories (*Carya*) and walnuts (*Juglans*). Reared on Shagbark Hickory (*C. ovata*) (TDS).

Notes. Moth rests head-down on tree trunks, usually from 3 to 8 ft. above ground; often on Shagbark Hickory (*C. ovata*), and sometimes partially hidden under "shags" (MCN, TDS).

C. *serena* W. H. Edwards (1864)
SERENE UNDERWING
Plate 1: 10

Description. FW dull gray, with little contrast; thin black lines, less dentate than in most species; subreniform broadly open. Collar on thorax brownish and contrasting. HW dull yellow-orange with black bands. Underside dull cream-orange with broad blackish bands; dusky over-all appearance. 50–60 mm. **H** 33: 14. **B** & **Mc** 7: 6.

Similar species. See *habilis*.

Range. Southern Massachusetts to North Carolina and west throughout area. May have ranged further north in the past and may presently be moving northward again (see Barnes & McDunnough [1918] and Sargent & Hessel [1970]).

Status. Uncommon to common. Apparently subject to long-term fluctuations in abundance.

Season. Mid-season to late (very long season). (Southern New England: 11 August; 18 July–6 October.)

Larval foodplants. Presumably hickories (*Carya*) and walnuts (*Juglans*).

Notes. Moth rests head-down, often on hickories (WAM).

A
B
C
D

Figure 2.4 The undersides of several black-winged species: (A) *C. andromedae;* (B) *C. miranda;* (C) *C. epione;* (D) *C. judith;* (E) *C. robinsoni;* (F) *C. lacrymosa;* (G) *C. dejecta;* (H) *C. insolabilis.* Many of these species are most easily identified on the basis of underside characteristics. 0.8X.

C. *robinsoni* Grote (1872)
ROBINSON'S UNDERWING
Plate 2: 3, 4

Description. FW light gray, much like *habilis;* but less contrast, and without prominent anal dash. Female, slightly more mottled, and with weak basal dash, is "curvata" French. HW black with contrasting white fringe. Underside boldly patterned in black and white, bands prominent (fig. 2.4E). Specimens with heavy blackish shading, running obliquely across FW, are "missouriensis" Schwartz; these are usually larger individuals, and may represent a distinct species (AEB). 60–70 mm. **H** 31: 7. **B** & **Mc** 2: 9, 10.

Similar species. Contrasting white fringe of HW will distinguish from most similar species (e.g., *angusi*). Some *flebilis* may be similar to "missouriensis," but former is generally smaller, with a much narrower pm white band on HW underside, and much less distinct white bands on FW underside.

Range. Southern Canada (Ontario), New Hampshire to West Virginia and Alabama, and west throughout area; "missouriensis" seems much more restricted (Pennsylvania to Florida and westward).

Status. Generally rare, especially to the north.

Season. Late. Perhaps the latest species in most areas.

Larval foodplants. Hickories (*Carya*), and perhaps walnuts (*Juglans*). Rearings especially important, particularly to determine the status of "missouriensis."

C. *judith* Strecker (1874)
JUDITH'S UNDERWING
Plate 1: 9

Description. FW uniform light gray, with only slight darkening about reniform and beneath anal dash. HW black; with dusky fringe, hardly contrasting. Underside, except for white basal areas, mostly blackish; bands indistinct (fig. 2.4D). 45–50 mm. **H** 32: 2. **B** & **Mc** 1: 15.

Similar species. Small size distinguishes from most other black-winged species, except *miranda* (which has distinct, though not prominent, white HW apex, and is entirely dusky beneath without white basal area) and *andromedae* (which is shaded blackish on inner FW margin, and has prominent white HW apex).

Range. Southern Canada (Ontario), New Hampshire to North Carolina and west throughout area.

Status. Generally uncommon.

Season. Mid-season. (Southern New England: 5 August; 21 July–26 September.)

Larval foodplants. Hickories (*Carya*) and walnuts (*Juglans*).

Notes. Seems to be a late flyer, taken more frequently in all-night light-traps than at bait or lights before 1:00 A.M. (see Appendix One). Moth rests head-down, often on hickories.

C. flebilis Grote (1872)
MOURNING UNDERWING
Plate 2: 8

Description. FW gray; shaded with brown, particularly in reniform; with heavy blackish shade running obliquely from basal dash to outer margin, just beneath apex (interrupted only near subreniform). HW black, with white fringe. 55–65 mm. **H** 32: 5 (as *carolina*, an atypical specimen). **B & Mc** 2: 12.

Similar species. See *robinsoni* "missouriensis." The HW fringe is blackish in *angusi* "lucetta."

Range. Massachusetts to North Carolina and west throughout area.

Status. Generally uncommon to rare

Season. Mid-season to late. (Southern New England: 30 August; 2 August–9 October.)

Larval foodplants. Hickories (*Carya*) (AEB).

Notes. Like *judith*, seems to be a late flyer (see Appendix One), and may therefore be considered very rare by collectors whose efforts terminate before the early-morning hours.

C. angusi Grote (1876)
ANGUS' UNDERWING
Plate 2: 5,6

Description. FW light gray, generally similar to *habilis*. Female, more mottled, and with prominent basal dash, is "edna" Beutenmüller. HW black, with mostly blackish fringe. Underside like *flebilis*. In form "lucetta" French, there is a heavy blackish shade running obliquely across FW (as in *flebilis* and "missouriensis." There is also a deep brown-black melanic form (**B & Mc** 2: 16). 60–70 mm. **H** 31: 11 "lucetta" as *flebilis*), 13. **B & Mc** 2: 13, 14 (female as form "edna"), 15 ("lucetta"), 16 (melanic).

Similar species. The blackish HW fringe will distinguish from most similar species. Some melanic *angusi* may approach *residua*, but underside differences (see *residua*) should permit distinction. The underside of *insolabilis* is always totally distinctive.

Range. Massachusetts to Georgia, and west and south throughout area. At present apparently very rare or absent in northern parts of range (recorded in Massachusetts and Connecticut in mid-1920s).

Status. Apparently subject to long-term fluctuations in abundance. Was widespread and common in 1890s and early decades of nineteenth century, but now rare or absent in the north, and generally uncommon in the south.

Season. Mid-season to late.

Larval foodplants. Hickories (*Carya*). Pecan (*Carya illinoensis*) in Texas (AEB).

C. *obscura* Strecker (1873)
OBSCURE UNDERWING
Plate 2: 1

Description. FW dull gray, much like *residua*, but generally even more uniform (i.e., with less contrastingly whitish st line; and usually without oblique dark streak extending to outer margin, just below apex, from near pm line). HW black, with largely whitish fringe. 60–70 mm. **H** (not shown). **B & Mc** 2: 17.

Similar species. Whitish HW fringe will usually distinguish from *residua*. These two species are easily separable in New England, but may intergrade to a greater extent to the south and west (see Barnes & McDunnough, 1918).

Range. Southern Canada (Ontario), Massachusetts to North Carolina and west throughout area.

Status. Generally uncommon. Almost always less common than *residua*.

Season. Late. (Southern New England: 6 September; 8 August–16 October.)

Larval foodplants. Hickories (*Carya*) and walnuts (*Juglans*). Reared on Pignut Hickory (*C. glabra*) (SAH) and Black Walnut (*J. nigra*) (DFS).

Notes. Moth rests head-down on tree trunks, usually from 3 to 8 ft. above ground; often on Shagbark Hickory (*C. ovata*), and sometimes partially or completely hidden under "shags" (WAM, MCN).

C. *residua* Grote (1874)
Plate 2: 2

Description. FW dull gray, like *obscura* (except as noted under that species). HW black, with mostly dusky fringe (except some white at apex). Underside without fuscous shading interrupting narrow white pm band on HW. 60–70 mm. **H** 31: 14 (as *obscura*). **B & Mc** 2: 18.

Similar species. See *obscura*. Some dark *angusi* may be similar, but these will usually have some fuscous shading interrupting narrow white pm band on HW underside.

Range. Southern Canada (Ontario), Maine to North Carolina and west throughout area.

Status. Subject to wide annual fluctuations in abundance, but generally common.

Season. Mid-season to late. (Southern New England: 19 August; 25 July–9 October.)

Larval foodplants. Hickories (*Carya*). Reared on Shellbark Hickory (*C. laciniosa*) (DFS) and Pignut Hickory (*C. glabra*) (SAH).

Notes. Moth rests head-down, often on Shagbark Hickory (*C. ovata*), and sometimes partially under "shags" (WAM).

C. sappho Strecker (1874)
SAPPHO UNDERWING
ERMINE UNDERWING
Plate 2: 11

Description. FW gray-white, rather uniform, except for blackening at costa near end of am line and above reniform; reniform itself red-brown. HW black with white fringe. 70–75 mm. **H** 31: 2. **B** & **Mc** 1: 14.

Similar species. FW completely distinctive.

Range. Gulf states, particularly Florida; straying northward to southern Illinois and Virginia.

Status. Generally considered very rare, but occurring regularly at specific locations in Florida and other Gulf states.

Season. Mid season.

Larval foodplant. Pecan (*Carya illinoensis*).

Notes. Moth rests head-down on light-barked trees; rather sluggish (JB).

C. agrippina Strecker (1874)
AGRIPPINA UNDERWING
Plate 2: 9

Description. FW dull gray-brown, with red-brown shading; particularly within am line, in st area, and in reniform. HW black with white fringe, latter distinctively barred at veins. Specimens with a dull greenish FW cast are "subviridis" Harvey. 75–85 mm. **H** 31: 1, 4 (incorrectly as "subviridis"). **B** & **Mc** 1: 1, 2, 3, 4 (aberration), 5, 6 ("subviridis").

Similar species. Combination of red-brown shadings on FW and white, heavily barred, HW fringe is quite distinctive. Some *retecta* "luctuosa" may be similar, but these will have rather heavy basal dash.

Range. Like *sappho*; Gulf states, straying northward to southern Illinois and New Jersey.

Status. Uncommon to common in Gulf states, rare and sporadic elsewhere.

Season. Mid-season to late.

Larval foodplant. Pecan (*Carya illinoensis*) (AEB).

C. *retecta* Grote (1872)
YELLOW-GRAY UNDERWING[1]
Plate 3 : 1, 2

Description. FW light gray, with accented pattern of black lines; basal dash usually prominent in both sexes, and succeeded by a similar dash, slightly higher, through am line; often with prominent anal dash as well; usually prominent blackish arc from costa, *under* reniform to outer margin below apex. HW black, with white fringe, lightly barred at veins. Underside boldly patterned in black and white, with pm white band prominent on both FW and HW. Large specimens with yellow-reddish FW cast are "luctuosa" Hulst, and may represent a distinct species. 60–70 mm. **H** 31: 8 ("luctuosa"). **B & Mc** 2: 11, 19 ("luctuosa"), 20 ("luctuosa").

Similar species. C. *dejecta* has a less prominent basal dash (usually absent in males) and lacks the succeeding dash through the am line; *dejecta* also has a narrower white HW fringe, and a less complete pm white band on FW underside. C. *vidua* is generally larger, with a much heavier am line, and the blackish FW arc extends *above* (not under) the reniform.

Range. Southern Canada (Ontario), Maine to Georgia and west throughout area. The form (possibly species) "luctuosa" seems confined at present to the more western and southern limits of this range.

Status. Common to abundant; except for "luctuosa," which seems to have become quite rare.

Season. Mid-season to late. (Southern New England: 9 September; 30 July–14 October.)

Larval foodplants. Hickories (*Carya*). Reared on Shagbark Hickory (C. *ovata*) (TDS).

Notes. Moth rests head-down on tree trunks, usually from 2 to 8 ft. above ground, rather often on hickories (MCN). Comes readily to baits.

[1] This common name is only appropriate for the form "luctuosa."

C. *ulalume* Strecker (1878)
Plate 3: 4

Description. FW like *dejecta* but "coarser," powdery or "dusted" appearance. FW ground less bluish than *dejecta*, and less brownish than *lacrymosa* (rather intermediate, neutral gray). HW black, with white fringe, barred as in *retecta* (often less barred than *lacrymosa*). Underside much like *dejecta*. 60–70 mm. **H** (not shown). **B & Mc** 2: 6 (as *lacrymosa*, variety *ulalume*; but somewhat atypical, particularly with regard to lack of HW barring).

Similar species. Has been consistently confused with *dejecta* and *lacrymosa*. The points noted above may aid in distinguishing, but positive identification is probably best left to experts.

Range. Difficult to specify because species may be overlooked or mis-identified. Known from Virginia, West Virginia, North Carolina, South Carolina, Georgia, and Mississippi in our area. AEB has studied extensively in Missouri (see Brower, 1922).

Status. Presumably very rare, but may be overlooked.

Season. Apparently mid-season.

Larval foodplants. Hickories (*Carya*) (AEB).

Notes. Moth very sluggish, entirely unlike *lacrymosa* (Brower, 1922).

C. *dejecta* Strecker (1880)

DEJECTED UNDERWING

Plate 3: 3

Description. FW gray, like *retecta,* but with less brownish tint and shadings, and without dash through am line. Pale area beyond am line along costa, extending obliquely toward subreniform, conspicuous. HW black, with white fringe (usually narrower than *retecta*). Underside similar to *retecta*, but with slightly more extensive and whiter basal area (fig. 2.4G). 60–70 mm. **H** 32: 1. **B** & **Mc** 2: 8 (atypically dark specimen).

Similar species. See *retecta* and *ulalume*.

Range. Massachusetts to Georgia and west throughout area.

Status. Generally rare and local.

Season. Mid-season, averaging much earlier than *retecta*. (Southern New England: 15 August; 19 July–18 September.)

Larval foodplants. Reared on Shagbark Hickory (*Carya ovata*) (TDS).

Notes. Moth rests head-down, usually on tree trunks, but occasionally under eaves and in other protected places on buildings, particularly during heavy rains.

C. *insolabilis* Guenée (1852)

INCONSOLABLE UNDERWING

Plate 2: 7

Description. FW gray, same shade as *dejecta;* rather uniform in male, but with strong contrasts and brown shading in female. Usually with contrasting blackish shading along inner margin of FW in both sexes. HW black with dusky fringe. Underside largely blackish, except whitish basal area; pm band on FW largely absent, and very narrow and obscure on HW (fig. 2.4H). 65–75 mm. **H** 31: 10. **B** & **Mc** 1: 7, 8.

Similar species. The dusky HW fringe and distinctive underside will distinguish this from similar species.

Range. Southern Canada (Ontario) to Florida and west throughout area. Northern records (New England) generally date back to the 1920s and 1930s.

Status. Rare to the north, uncommon elsewhere.

Season. Mid-season (long season).

Larval foodplants. Hickories (*Carya*) (AEB).

C. *vidua* (Smith & Abbot) (1797)
WIDOW UNDERWING
Plate 2: 10

Description. FW light gray, strongly marked with black lines and shadings; prominent blackish arc from *above* reniform at costa to outer margin, just below apex; heavy basal and anal dashes. HW black with broad white fringe. Underside boldly patterned in black and white, all bands prominent. 70–80 mm. **H** 31: 5. **B** & **Mc** 1: 17.

Similar species. See *retecta* and *maestosa*.

Range. Southern Canada (Ontario), Maine (rare) and New Hampshire to Florida and west throughout area.

Status. Generally common, but becoming rarer and more sporadic to the north.

Season. Late.

Larval foodplants. Walnuts (*Juglans*) and hickories (*Carya*).

Notes. Moth rests head-down on tree trunks, usually from 1 to 8 ft. above ground; often on hickories and oaks (WAM, MCN). Comes readily to baits.

C. *maestosa* (Hulst) (1884)
Plate 2: 12

Description. FW light gray, with lines as in *vidua*, but not strongly marked except for blackish arc from costa *above* reniform to just below apex at outer margin. HW black, with extensive gray basal hairs, and white fringe. Underside similar to *vidua*. The name "moderna" Grote was applied to a small, aberrant specimen, and the name should not be retained. 80–90 mm. **H** 31: 15 (as *viduata*). **B** & **Mc** 1: 20.

Similar species. Only *vidua* is vaguely similar, but it is usually smaller, has a more extensive black FW pattern, and a broader HW fringe.

Range. Long Island, New York (1933, SAH) to Florida and west throughout area.

Status. Generally uncommon; may be locally common in the south, but very rare to the north in its range. Seems periodically to become established as far north as New Jersey at present (DFS).

Season. Mid-season to late.

Larval foodplants. Walnuts (*Juglans*) and hickories (*Carya*), including Pecan (*C. illinoensis*).

C. lacrymosa Guenée (1852)
TEARFUL UNDERWING
Plate 3: 5, 6, 7

Description. FW dark gray, with prominent black lines, and brown and whitish shadings; brown particularly in st area. HW black, with prominently barred white fringe. Underside black and white, with relatively narrow pm white band on both FW and HW (compared to *palaeogama*) (figs. 2.4F; 5.5A). The species is very variable. Among the named forms are "evelina" French (inner margin and patch below apex of FW contrastingly blackish), "paulina" H. Edwards (blackish from base to pm line of FW), and "zelica" French (basal and pm-st area of FW contrastingly blackish). An aberrant specimen (plate 3: 8) with *lacrymosa*-like FW, and traces of orange on the HW, seems closest to this species, but shows some *palaeogama* characteristics, and may represent a hybrid between these two species (see chapter five). 60–70 mm. **H** 31: 6, 9 ("evelina"), 12 ("paulina"). **B & Mc** 2: 1, 2 ("zelica"), 3 ("evelina"), 4 ("paulina"), 5 ("paulina"), 7.

Similar species. FW remarkably similar to *palaeogama*, and with parallel variations. A species as variable as this will always provide some identification problems. Questionable individuals should be compared with long series of collected specimens. See *ulalume*.

Range. Massachusetts to Florida and west throughout area.

Status. Generally uncommon to common, but rare to the north and northeast in its range. Exhibits marked fluctuations in annual abundance.

Season. Mid-season to late.

Larval foodplants. Hickories (*Carya*).

Notes. Unlike many of the black-winged species, this moth is very skittish while resting, and may fly long distances when disturbed.

C. palaeogama Guenée (1852)
OLDWIFE UNDERWING
Plate 3: 9, 10, 11, 12

Description. FW like *lacrymosa*, with parallel variations. HW orange, including apex; with black bands, and heavy brown-black basal hairs. Underside orange and blackish, with prominent bands; body and basal areas whitish (fig. 5.5C). Among the forms of this very variable species are "annida" Fager (inner margin and patch below apex of FW contrastingly blackish), and "phalanga" Grote (basal and pm-st area of FW contrastingly blackish). The species also occurs in melanic forms, one of which may correspond to "denussa" Ehrman (though some authorities consider this a melanic of *habilis*, e.g., Forbes, 1954). An aberrant specimen with the HW almost completely black (plate 3: 8) may be this species, but it seems closer to *lacrymosa* in many respects, and is analyzed as a possible hybrid in chapter five. Some specimens, particularly females of "annida," are strongly mottled with white. 60–70 mm. **H** 36: 3, 4 ("phalanga"). **B & Mc** 6: 18, 19 ("phalanga"), 20 ("annida").

Similar species. As with *lacrymosa*, this species is best learned by studying long series of specimens. The shape and course of the inner black band on the HW is quite constant. *C. neogama* has the body and basal areas on underside fuscous, not whitish.

Range. Southern Canada (Ontario), Maine (rare) to North Carolina and west throughout area.

Status. Common to abundant. Subject to wide fluctuations in abundance from year to year.

Season. Mid-season to late. (Southern New England: 22 August; 16 July–9 October.)

Larval foodplants. Hickories (*Carya*) and walnuts (*Juglans*). Reared on Shagbark Hickory (*C. ovata*) (TDS).

Notes. Moth rests head-down on tree trunks, usually from 4 to 8 ft. above ground, often on hickories (MCN). Seems to be more commonly taken in all-night light-traps than otherwise, and so may be a late flyer.

C. *nebulosa* W. H. Edwards (1864)
CLOUDED UNDERWING
Plate 4: 1

Description. FW yellowish to reddish brown, much darker within am line, and somewhat darker at apex and anal angle. HW yellow, with black bands. 75-85 mm. **H** 33: 16. **B** & **Mc** 6: 17.

Similar species. Unmistakable.

Range. Long Island, New York, to Mississippi and west throughout area.

Status. Rare.

Season. Apparently mid-season to late.

Larval foodplant. Unknown.

Notes. Moth found resting head-down on tree trunks ((WAM), in or under root-tangles along stream banks (MCN), and under fallen tree trunks lying across ravines (JB).

C. *subnata* Grote (1864)
YOUTHFUL UNDERWING
Plate 4: 2

Description. FW pale gray, with thin blackish lines and light brown shadings. Female more mottled than male, with slight basal dash (usually completely absent in male). HW yellow-orange with black bands; fringe broad and often mostly yellow-orange (males); basal hairs only slightly darker than ground. Underside light yellow-orange and black; with very wide, light outer border and fringe on HW. 75-90 mm. **H** 33: 15. **B** & **Mc** 6: 15, 16.

Similar species. Often confused with *neogama*, but usually paler and less contrastingly marked on FW; HW has narrower black bands and less contrasting basal hairs, giving a much paler overall appearance. Male *neogama* generally have at least a trace of a basal dash.

The following distinction between these two species was discovered by Douglas C. Ferguson, and seems entirely reliable. It is best seen under magnification.

C. *subnata* — hind tibia almost cylindrical in cross-section, densely and uniformly covered with spines on outer (ventral) surface.

C. *neogama* — hind tibia strongly compressed in cross-section, sparsely and sporadically spined on outer (ventral) surface.

Range. Southern Canada (Ontario, Nova Scotia — one record) to North Carolina and west throughout area.

Status. Generally quite rare, but apparently exhibits long-term fluctuations in abundance (Forbes, 1954).

Season. Mid-season. (Southern New England: 10 August; 26 July-9 September.)

Larval foodplants. Butternut (*Juglans cinerea*) (JGF), and possibly Black Walnut (*J. nigra*) and hickories (*Carya*).

Notes. Moth rests head-down on tree trunks, often hickories and walnuts (MCN).

C. neogama (Smith & Abbot) (1797)
THE BRIDE
Plate 4: 3, 4

Description. FW dull gray (dark specimens) or light gray-brown (light specimens); often with considerable brown in am, st, and reniform areas. Light specimens are usually males, and show at least a trace of the basal dash; dark specimens are usually females, and show a short, heavy basal dash. HW deep yellow-orange with black bands; dense, brown basal hairs. Underside much like *subnata*. The melanic form, with entirely black FW, is "mildredae" Franclemont. 70–85 mm. **H** 36: 5. **B** & **Mc** 6: 10, 11, 12.
Similar species. See *subnata*.
Range. Southern Canada (Ontario, Quebec) to Georgia and west throughout area.
Status. Uncommon to common.
Season. Mid-season to late (averaging considerably later than *subnata*). (Southern New England: 8 September; 19 July–14 October.)
Larval foodplants. Hickories (*Carya*) and walnuts (*Juglans*). Reared on butternut (*J. cinerea*) (AEB).
Notes. Moth rests head-down on tree trunks, often on hickories (WAM).

C. ilia (Cramer) (1775)
ILIA UNDERWING THE WIFE
THE BELOVED UNDERWING
Plate 4: 5, 6, 7, 8

Description. Highly variable. FW generally gray, mottled with black and white; reniform typically outlined with white. HW orange-red with black bands; fringe cream, with some orange, and heavily barred. Among the named forms and varieties of this species are "confusa" Worthington (FW concolorous gray-brown), "umbrosa" Worthington (no white in or around reniform), "conspicua" Worthington (reniform solid white), "normani" Bartsch ("semi-melanic," with blackish FW from base to pm line, and extended black on HW), and "satanas" Reiff (melanic, with entirely black FW). 70–80 mm. **H** 34: 7 (as *osculata*), 14 ("umbrosa"), 17 ("conspicua," as *uxor*). **B** & **Mc** 6: 4 ("umbrosa"), 5 ("confusa"), 6, 7 ("conspicua").
Similar species. Because of its great variability, *ilia* is frequently mis-identified. Note that the HW color and pattern varies far less than the FW (as usual in the *Catocala*). Questionable individuals should be studied with reference to a long series of collected specimens. See *innubens*.
Range. Southern Canada (Ontario, Quebec, Nova Scotia) to Florida and west throughout area.
Status. Common to abundant.
Season. Early, but some individuals persisting until late. These late individuals have a mottled appearance, often with whitish patches presumably resulting from wear. (Southern New England: 2 August; 10 July–8 October.)
Larval foodplants. Oaks (*Quercus*). Reared on Black Oak (*Q. velutina*) (TDS), Red Oak (*Q. borealis*) (AEB), and White Oak (*Q. alba*) (AEB).
Notes. Moth rests head-down on tree trunks, often high, and frequently on oaks (WAM, TDS). Comes readily to baits.

C. cerogama Guenée (1852)
YELLOW-BANDED UNDERWING
Plate 5: 1, 2

Description. FW light gray, extensively shaded with brown; pale patch after am line along costa, and extending obliquely inward to the very pale subreniform. HW yellow with very broad inner black band; brown basal hairs completely filling basal area; pm yellow band very even, and followed by very broad outer black band; fringe light yellow and cream, heavily barred. Underside pale whitish-yellow, with all bands prominent. Among the named varieties of this species are the forms "bunkeri" Grote (with heavy brown FW shadings) (fig. 5.4), and "eliza" Fischer (with white FW patches), and the aberration "aurella" Fischer (with yellow basal hairs on HW). The melanic form is "ruperti" Franclemont, and has a deep brown-black FW. 70–80 mm. **H** 34: 6. **B & Mc** 6: 1.

Similar species. The unusual HW renders the species generally unmistakable.

Range. Southern Canada (Ontario, Nova Scotia) to North Carolina (Fontana) and west throughout area.

Status. Common where foodplant occurs.

Season. Mid-season to late.

Larval foodplants. Basswood (Linden) (*Tilia americana*). There may be another foodplant, as the species is sometimes found far from Basswood.

Notes. Moth rests head-down, often high, on large, light trees (particularly White Oak) (WAM). Comes readily to baits.

C. relicta Walker (1857)
FORSAKEN UNDERWING THE RELICT WHITE UNDERWING
Plate 4: 9, 10, 11, 12

Description. FW white, variously marked with gray and blackish lines and patches. HW black with even white pm band and white fringe. Underside boldly patterned in black and white, all bands prominent, and pronounced discal lunule on HW. Typical specimens have the basal and st areas of the FW largely filled with blackish, while form "clara" Beutenmüller has these areas largely whitish. Form "phrynia" H. Edwards is evenly dusted with grayish over the entire FW. 70–80 mm. **H** 32: 6 ("clara"), 7 (typical, as variety "bianca"). **B & Mc** 1: 9, 10 ("phrynia"), 11 ("clara"), 12, 13.

Similar species. Unmistakable.

Range. Hudson Bay area to Kentucky and west throughout area.

Status. Common to the north, and becoming very rare to the south, in this range. Subject to considerable fluctuation in annual abundance.

Season. Mid-season. (Southern New England: 12 August; 19 July–4 October.)

Larval foodplants. Poplars (*Populus*). Reared on Trembling Aspen (*P. tremuloides*) and Lombardy Poplar (*P. nigra italica*) (TDS). May also take some willows (*Salix*).

Notes. Moth rests head-up, particularly on trunks of light trees (primarily White Birch, *Betula papyrifera*, in southern New England; also aspens and lighter oaks and maples); prefers white backgrounds in experimental tests (Sargent, 1973*a*). Easily mated in captivity (Sargent, 1972*a*). Comes readily to baits.

C. marmorata W. H. Edwards (1864)
MARBLED UNDERWING
Plate 5: 3

Description. FW light gray, with brownish and white shadings, latter particularly in st area; blackish arc from middle of costa down through reniform and over to outer margin just below apex. HW pinkish-red with black bands and white, scalloped fringe. 90–95 mm. **H** 35: 9. **B** & **Mc** 3: 19.

Similar species. Large size, pink-red HW, and prominent blackish FW arc will distinguish from any other species.

Range. Vermont (old records); Long Island, New York (SAH, 1930–32) to South Carolina and west throughout area; particularly in mid-Atlantic and southern Appalachian states at present.

Status. Rare.

Season. Long season, but mostly midseason.

Larval foodplants. Unknown.

Notes. Moth rests head-down on tree trunks, often wedged into crevices; rather sluggish (JB).

C. unijuga Walker (1857)
ONCE-MARRIED UNDERWING
Plate 5: 5, 6

Description. FW gray, mottled with whitish, and with st line regularly dentate and contrastingly white. HW orange-red to scarlet-red with black bands; white, heavily barred fringe. Underside with white FW ground, white and reddish HW ground, all bands prominent; usually with well-marked discal lunule. The melanic form, "agatha" Beutenmüller, is dark, smoky gray, but with white st line usually contrasting. An aberration with HW entirely black is "fletcheri" Beutenmüller. 70–90 mm. **H** 33: 5. **B** & **Mc** 1: 19 ("fletcheri"); 4: 6, 7 ("agatha"); 8: 23 ("beaniana").

Similar species. Close to *meskei*, but that species usually smaller, with FW giving "dusty" impression. HW in *meskei* is slightly paler, tending toward salmon; with inner black band narrower, and tapering more gradually to point at or before inner margin; apex often with some orange. The discal lunule on HW underside is usually very faint or absent in *meskei*.

See *semirelicta*.

Range. Southern Canada (Ontario, Quebec, Nova Scotia) to Pennsylvania and west throughout area.

Status. Common over most of range.

Season. Early to late, perhaps the longest season of any *Catocala*. (Southern New England: 28 August; 8 July–11 October.)

Larval foodplants. Poplars (*Populus*) and willows (*Salix*). Reared on Trembling Aspen (*P. tremuloides*) (TDS).

Notes. Moth rests head-up on tree trunks, often high. Captured feeding at night at milkweed and joe-pye-weed (MCN). Comes readily to baits.

C. parta Guenée (1852)
MOTHER UNDERWING
Plate 5: 11, 12

Description. FW light gray, shaded with light brown and white; with distinct dark dashes in areas of basal and anal dashes, and obliquely from pm line to outer margin just below apex. HW salmon-red with black bands. Underside HW with distinct, black discal lunule. More contrasting specimens, with the inner margin of FW darkened, are "perplexa" Strecker. The melanic form, with blackish FW, is "forbesi" Franclemont. An aberration with yellow HW ground, which may simply be due to fading, has been named "petulans" Hulst. 75–85 mm. **H** 34: 11. **B** & **Mc** 3: 14.

Similar species. C. coccinata shows a somewhat similar FW pattern, but HW shade (crimson) is entirely different in that case. *C. marmorata* is similar in FW and HW colors, but is larger and has distinctive blackish arc on FW.

Range. Hudson Bay area to New Jersey and Pennsylvania (perhaps further in mountains) and west throughout area. A distinctly northern species.

Status. Uncommon to common.

Season. Late. (Southern New England: 12 September; 29 July–14 October.)

Larval foodplants. Willows (*Salix*) and poplars (*Populus*). Reared on Black Willow (*S. nigra*) (DFS).

Notes. Moth rests head-up on tree trunks, usually from 4 to 8 ft. above ground, often on White Oak; quite skittish (WAM, MCN).

C. briseis W. H. Edwards (1864)
BRISEIS UNDERWING
Plate 5: 7, 8

Description. FW grayish black, dusted with white; usually with whitish patch near (and sometimes including) subreniform; area between pm and st lines contrastingly brownish and/or whitish, and with distinctive vertical ribbing of the scales under magnification. HW scarlet with black bands. Underside like *parta*, but with darker FW and HW margins. Specimens with light gray FW are "albida" Beutenmüller. 60–70 mm. **H** 32: 4 (as *groteiana*); 35: 12. **B** & **Mc** 3: 5, 6, 8 ("albida").

Similar species. Unmistakable when pm-st area patchily filled with whitish (similar then to the western *groteiana*). Specimens without this whitish might be confused with *unijuga* or *meskei*, but these usually have paler FW, less deeply scarlet HW, and more irregular HW bands.

Range. Hudson Bay area to at least New Jersey and Pennsylvania, and west throughout area. A distinctly northern species.

Status. Generally uncommon to rare, but occasionally common. Apparently rare in many places at present, but probably subject to long-term fluctuations in abundance.

Season: Mid-season to late. (Southern New England: 20 August; 26 July–22 September.)

Larval foodplants. Poplars (*Populus*) and willows (*Salix*). Reared on Trembling Aspen (*P. tremuloides*) (AEB).

C. *semirelicta* Grote (1874)
Plate 5: 9, 10

Description. FW whitish, shaded with gray; lines black and contrasting; pm and st lines regularly dentate, with whitish area between; st immediately followed by very prominent, regularly dentate, black line; usually a blackish shade extending horizontally from basal dash to outer margin (this shade absent in more uniform specimens, which are "atala" Cassino). HW pinkish red with black bands; inner band usually terminating well before inner margin. 65–75 mm. **H** (not shown). **B** & **Mc** 4: 2.

Similar species. C. *unijuga* may be quite similar, but that species is usually larger, has less contrastingly black FW lines, particularly beyond the white st, and a heavier inner black band on HW which usually reaches the inner margin.

Range. Quebec, Nova Scotia, Maine, and westward through Ontario and Manitoba. Ranges less far south than any other eastern *Catocala*.

Status. Uncommon to rare.

Season. Very long season, but predominantly mid-season.

Larval foodplants. Poplars (*Populus*), possibly only Balsam Poplar (*P. balsamifera*) (DCF).

C. *meskei* Grote (1873)
MESKE'S UNDERWING
Plate 5: 4

Description. Very similar to *unijuga*. FW gray, with whitish and blackish dusting, giving a more uniform appearance than *unijuga*. HW orange-red with black bands; white apex often with orange dots (unlike *unijuga*). Underside HW with discal lunule faint or absent. The melanic form, with uniformly blackish FW, is "krombeini" Franclemont. 65–75 mm. **H** 33: 6 (atypically pale; possibly faded). **B** & **Mc** 4: 8.

Similar species. Unless as distinctive as the specimen on plate 5, *meskei* is one of the most difficult of the eastern *Catocala* to identify. Positive identification is probably best left to experts, but see *unijuga* for aids to tentative identification.

Range. Southern Canada (Quebec, Ontario) to Pennsylvania and west throughout area. Apparently very sporadic in occurrence within this range.

Status. Uncommon to rare. Appears to be more common at present in the western portions of its range.

Season. Mid-season, averaging earlier than *unijuga*.

Larval foodplants. Poplars (*Populus*), particularly Cottonwood (*P. deltoides*), and willows (*Salix*). Reared on Large-tooth Aspen (*P. grandidentata*) (SAH).

Notes. Moth rests head-up on tree trunks (WAM).

C. *junctura* Walker (1857)
Plate 6: 1

Description. FW rather uniform gray, tinted and shaded with brown; lines somewhat blurred, giving "soft" appearance. HW scarlet-orange with black bands; inner black band rather narrow and sometimes broken before inner margin; apex often heavily shaded with orange. Underside without HW discal lunule. Specimens with a dark shade running across FW from basal dash to near outer margin are "julietta" French. 70–75 mm. **H** (not shown). **B** & **Mc** 8: 6, 7.

Similar species. Somewhat similar to *unijuga*, but the soft brownish-gray FW, and very different HW banding pattern, are distinctive.

Range. Western New York to Pennsylvania and west and south throughout area.

Status. Uncommon to common.

Season. Very long season; early to quite late.

Larval foodplants. Willows (*Salix*).

Notes. Moth tends to rest in caves by day (see Brower, 1930).

C. *cara* Guenée (1852)
DARLING UNDERWING
BRONZE UNDERWING
Plate 6: 2

Description. FW even gray-brown, with greenish dusting and shadings; am and pm lines black, fine and single, though sometimes obsolete. HW deep pink with broad and even black bands; inner black band rather erect and reaching inner margin; dirty white fringe, without distinct apex. Underside dirty white with blackish bands (fig. 5.6C). Specimens with a pale apical patch on FW are "carissima" Hulst, which is predominantly southern but has been taken as far north as Long Island, New York (1929, 1933, SAH). 70–85 mm. **H** 32: 9. **B** & **Mc** 3: 9, 10 ("carissima").

Similar species. Essentially unmistakable, though occasional intergrades to *amatrix* may be confusing (see chapter five on hybrids). Ordinarily, *amatrix* has more contrasting am and pm lines on FW, and a distinctive HW color and banding pattern.

Range. Canada (Ontario) (not definitely known from Maine) to Florida (form "carissima") and west throughout area.

Status. Common.

Season. Mid-season to late. (Southern New England: 6 September; 2 August–27 October.)

Larval foodplants. Willows (*Salix*), particularly Black Willow (*S. nigra*), and poplars (*Populus*).

Notes. Moth often rests in protected places, e.g., in caves, under eaves, bridges, etc. When on trees, rests head-down, usually from 3 to 5 ft. above ground, and is often very skittish. Comes readily to baits.

C. concumbens Walker (1857)
SLEEPY UNDERWING
PINK UNDERWING
Plate 6: 5

Description. FW uniform light silvery gray, becoming whitish toward costa; am and pm lines black, fine and single. HW pink, with broad and even black bands, and pure white fringe. Aberrational specimens include those with conspicuous pink hairs on the abdomen ("diantha" H. Edwards), and with yellow HW ground ("hilli" Grote). Occasional specimens have the HW bands somewhat blurred (plate V). 60–75 mm. **H** 35: 10. **B & Mc** 3: 15.

Similar species. Combination of silvery gray FW, and pink, evenly banded HW is distinctive.

Range. Southern Canada (Ontario, Quebec, Nova Scotia) to North Carolina (Highlands) and west throughout area.

Status. Generally common, but becoming rare to the south in its range.

Season. Mid-season. (Southern New England: 20 August; 19 July–9 October.)

Larval foodplants. Willows (*Salix*) and poplars (*Populus*). Apparently a "fussy" feeder in some instances, perhaps locally requiring specific host species (DFS, TDS). Reared on Trembling Aspen (*P. tremuloides*) (AEB).

Notes. Moth rests head-down, often low, on tree trunks; also on fence posts, rock surfaces, etc. Tends to drop to ground when disturbed (WAM). Comes readily to baits.

C. amatrix (Hübner) (1818)
THE SWEETHEART
Plate 6: 3, 6, 8

Description. FW dull gray-brown; with am and pm lines black and contrasting on upper half, lighter on lower half, subreniform large and only slightly separated from am line by a narrow blackish patch; basal dash blackish and heavy, the blackish often continuing obliquely across FW to near apex. In form "selecta" Walker, without these blackish shadings. HW pinkish-red with black bands; the inner band further out and more excurved than in *cara*. Underside with whitish FW ground (paler than *cara*) (fig. 5.6A). Occasional hybrids with *cara* may occur (see chapter five). 75–85 mm. **H** 32: 12 ("selecta"), 13 (as variety "nurus"). **B & Mc** 3: 11, 12 ("selecta").

The melanic specimen on plate 6:3 seems to be this species. Note the course and contrast of the am and pm lines; the whitish, slightly contrasting, st line; and the color and banding pattern of the HW — all of which fit *amatrix*, rather than *cara*. I would suggest applying the name "hesseli" to this beautiful form, in honor of Sidney A. Hessel, who took two such specimens at Woodmere, New York (21 and 28 August 1932)

Similar species. See *cara*.

Range. Southern Canada (Ontario), Maine (rare) to Florida and west throughout area.

Status. Generally uncommon to rare, but sometimes locally common.

Season. Mid-season to late (almost identical to *cara*). (Southern New England: 6 September; 5 August–10 October [CGK].)

Larval foodplants. Willows (*Salix*) and poplars (*Populus*). Apparently rather restricted, perhaps most often on Black Willow (*S. nigra*) and Cottonwood (*P. deltoides*).

Notes. Like *cara*, moth often rests in protected places, e.g., under loose bark, in caves, under bridges, and particularly on buildings. Very skittish, often flying long distances if disturbed. Comes readily to baits.

C. *delilah* Strecker (1874)
DELILAH UNDERWING
Plate 7: 1

Description. FW dull gray-brown, with ligher shadings and conspicuous blackish lines; area between pm and st lines brownish. HW yellow-orange, with black bands; apex and fringe entirely yellow-orange. Underside deep yellow with black bands. 60-65 mm. **H** 34: 4. **B** & **Mc** 6: 13, 14 (pale, western "desdomona").

Similar species. Vaguely similar to *consors*, but that species with dirty whitish HW fringe, and entirely distinctive (zigzag) HW banding pattern. See *muliercula*.

Range. Just barely within western edge of our area. Southern Illinois to Mississippi and west. Florida specimens (Kimball, 1965) may be strays.

Status. Absent from most of our area; otherwise uncommon to rare (more common west of our area).

Season. Mid-season.

Larval foodplants. Oaks (*Quercus*).

C. *illecta* Walker (1857)
MAGDALEN UNDERWING[2]
Plate 7: 2

Description. FW similar to *concumbens*; even, pale gray; with fine, black, single lines; st blurred whitish. HW light yellow with widely separated black bands; inner band irregular and ending abruptly before inner margin. 60-70 mm. **H** (not shown). **B** & **Mc** 7: 13.

Similar species. Quite distinctive. *C. clintoni* is smaller, with prominent basal dash.

Range. Southwestern Ontario to South Carolina, and west and south (to Texas).

Status. Generally uncommon to rare, becoming more common to the west of our area.

Season. Early.

Larval foodplants. Honey Locust (*Gleditchia triacanthos*). Possibly also on Lead Plant (*Amorpha*).

[2] This common name is based on a synonym for *illecta*, *magdalena* Strecker (1874).

C. *abbreviatella* Grote (1872)
Plate 7: 3

Description. FW uniform light brown-gray; lines black, becoming obsolete below middle of FW; reniform with black outer ring. HW yellow-orange with black bands; inner band ending abruptly before inner margin. 40–50 mm. **H** 34: 9. **B & Mc** 10: 18.

Similar species. This moth is part of an essentially western complex of Lead Plant (*Amorpha*) and Locust (*Robinia*) feeders. *C. nuptialis* (plate 7: 6) is closely similar to *abbreviatella,* but has a small, solid black reniform. *C. whitneyi* (plate 7: 7) is also similar, but has a dark, heavy am line (ending in a broad triangle at the middle of FW), and a blackish patch around reniform. *C. amestris* has doubled FW lines. See *clintoni*.

Range. Essentially western, but with sporadic eastern records. *C. abbreviatella* from Manitoba to Florida and westward (also reported from Pennsylvania). *C. nuptialis* at extreme west of our area (Illinois, Wisconsin) and westward. *C. whitneyi* from Ohio and Tennessee, west and north to Kansas and Manitoba.

Status. Sporadic, rare, or accidental over most of our area.

Season. Early, with *whitneyi* apparently somewhat later than the others.

Larval foodplants. Lead Plant (*Amorpha*).

C. *amestris* Strecker (1874)
Plate 7: 4

Description. FW like *abbreviatella*, but with stronger lines and shadings; am and pm lines clearly double, at least halfway across FW; reniform with double black ring, somewhat obscured by blackish shading. HW yellow-orange (deeper than *abbreviatella* and others in that group), with black bands; outer black band broken, or complete ("westcotti" Grote); inner black band more erect and sharply angled than in *abbreviatella*. 45–50 mm. **H** (not shown). **B & Mc** 8: 17, 18 ("westcotti").

Similar species. See *abbreviatella*.

Range. Common in Texas; otherwise rare and sporadic. Recorded from North and South Carolina and Florida; and Wisconsin and Illinois at the western edge of our area.

Status. Generally rare and sporadic in the east, though regularly occurring at some locations (e.g., McClellanville, South Carolina) (RBD).

Season. Early.

Larval foodplants. Lead Plant (*Amorpha*) and Locust (*Robinia*).

C. *messalina* Guenée (1852)
Plate 7: 8

Description. FW uniform dull gray, darkening toward outer margin; all lines and markings obsolete, or nearly so. HW light yellow-orange, with broad blackish outer band (no inner band), and wide whitish-yellow fringe. Underside yellow-cream and fuscous, with pm black band on FW, and trace of inner black band on HW toward costa. 40–45 mm. **H** 36: 1. **B** & **Mc** 10: 20.

Similar species. The other *Catocala* with only the outer band on HW upperside, *amica* and *jair*, have prominent FW markings.

Range. Southwestern, particularly Texas. Recorded from Florida, South Carolina and Virginia.

Status. Very rare and local in the east.

Season. Apparently early to mid-season.

Larval foodplants. Unknown.

C. *sordida* Grote (1877)
Plate 7: 12, 13, 15, 16

Description. FW gray, somewhat mottled with whitish and blackish shadings; no basal dash; inner margin sometimes narrowly blackish (form "metalomus" Mayfield). HW yellow-orange, with black bands; outer band broken, with separate anal spot. HW underside with blackish shading tending to complete the inner black band to inner margin basally. The melanic form, "engelhardti" Lemmer, may be very difficult to separate from the melanic form of *gracilis*. 40–45 mm. **H** 35: 7 (as *praeclara*). **B** & **Mc** 9: 8, 9 (both as *gracilis*, form "sordida").

Similar species. Very close to *gracilis*, with some authors suggesting the two may be identical (e.g., Barnes & McDunnough, 1918; Adams & Bertoni, 1968). Generally, *sordida* is distinguished by a less mottled FW overall, less extensive blackish shading along the inner margin, and absence of a basal dash. In addition, the inner black band on the HW underside tends to form a complete loop, rather than ending abruptly as in *gracilis*. These characters are not always correlated, and many collectors finally resort to separating the two species on the basis of the presence (*gracilis*) or absence (*sordida*) of the FW basal dash.

Range. Southern Canada (Manitoba to Nova Scotia) to Georgia and west throughout area.

Status. Common. Varies from considerably more common to considerably less common than *gracilis* from place to place (and even season to season in any one place).

Season. Mid-season. (Southern New England: 4 August; 5 July–13 September. Consistently averaging earlier than *gracilis*.)

Larval foodplants. Blueberries, and perhaps related plants (*Vaccinium*). May have different foodplant species than *gracilis*.

Notes. Moth rests head-down on tree trunks, usually from 4 to 8 ft. above ground. Apparently prefers rough-barked trees, often resting deep in furrows.

C. *gracilis* W. H. Edwards (1864)
GRACEFUL UNDERWING
Plate 7: 14, 17, 19, 20

Description. Similar to *sordida*, except as noted under that species. FW gray, mottled with whitish and blackish shadings; generally with rather extensive blackish shading along inner margin (sharply set off, and running across thorax, in form "lemmeri" Mayfield; absent in form "cinerea" Mayfield); distinct basal dash; lines fine, black (sometimes broken or obsolete); st usually distinct. HW as in *sordida*. Underside HW with inner black band tending to end abruptly after angling back toward inner margin. Melanic individuals occur, but these may be difficult to distinguish from melanic *sordida*. Melanic specimens with a distinct basal dash (plate 7: 19) are presumably this species, and I would suggest the name "hesperis" for this form, which has not been previously named. 40–45 mm. **H** 35: 8. **B** & **Mc** 9: 7.

Similar species. See *sordida* and *louiseae*.

Range. Southern Canada (Manitoba), Maine to Florida (rare) and west throughout area.

Status. Common most places throughout range.

Season. Mid-season. (Southern New England: 12 August; 18 July–14 September. Consistently averaging later than *sordida*.)

Larval foodplants. Blueberries, and perhaps related plants (*Vaccinium*). Reared on Late, or Half-high, Blueberry (*V. vacillans*) (TDS).

Notes. Moth rests as *sordida*, often in furrows of large, rough-barked trees.

C. *louiseae* Bauer (1965)
Plate 7: 18

Description. FW similar to C. *andromedae*, with prominent white line immediately following black pm line, small black spot *on* am line at level of subreniform (not *between* am line and subreniform as in *andromedae*), and blackish shadings along inner margin (broadening between pm line and outer margin) and near apex. HW deep yellow-orange (as in *amestris*) with black bands; inner band ending abruptly before inner margin. 40 mm.

Similar species. The *andromedae*-like FW, with yellow-orange and black banded HW, is quite distinctive. However, this moth is very close to *gracilis*. I recently discovered two older specimens of *louiseae* at the U. S. National Museum (from Arkansas and Florida) which had been placed with *gracilis*. Perhaps *louiseae* is a southern subspecies of *gracilis*. Rearings should be carried out to establish the status of this moth.

Range. Described from St. Johns Co., Florida.

Status. Presumably rare, or very local.

Season. Early (All specimens mentioned by Bauer were taken in May.)

C. *andromedae* (Guenée) (1852)
GLOOMY UNDERWING
Plate 7: 21

Description. FW like *gracilis*, but with more contrasting whitish shadings, particularly between pm and st lines; blackish shading along inner margin always present, and usually heavy; prominent black spot between am line and subreniform. HW black, with white apex and blackish fringe. Underside blackish, with only a broad pm band on FW, and apex of both FW and HW contrastingly white (fig. 2.4A). 40–50 mm. **H** 32: 3 (as *tristis*). **B** & **Mc** 1: 18.

Similar species. The smallest black-winged *Catocala*. Prominently marked FW, particularly blackish inner margin, easily distinguishes *andromedae* from *miranda* and *judith*. *C. louiseae*, with similar FW, has yellow and black-banded HW.

Range. Southern Canada (Ontario, Quebec), Maine to Florida (rare) and west throughout area.

Status. Uncommon to common, but somewhat sporadic. Apparently this species, like others of the *Vaccinium* feeders, is more common toward the coast than inland.

Season. Mid-season. (Southern New England: 10 August; 8 July–10 September. Overlapping both *gracilis* and *sordida*; though generally, like *gracilis*, averaging a little later than *sordida*.)

Larval foodplants. Blueberries (*Vaccinium*) and possibly Bog Rosemary (*Andromeda*). Reared on Low Blueberry (*V. pennsylvanicum*) (TDS).

Notes. Resting habits of moth essentially identical to *gracilis*.

C. *herodias* Strecker (1876)
HERODIAS UNDERWING
Plate 6: 10

Description. FW rather uniform gray; with whitish, brownish, and blackish shadings; veins blackish and somewhat contrasting, alternating with parallel whitish lines toward outer margin. All eastern specimens are race *gerhardi* Barnes & Benjamin, having the FW costa broadly shaded with whitish-gray (typical specimens, from Texas, have the FW costa concolorous). HW bright crimson-red with black bands. Underside like *coccinata*. 55–65 mm. **H** (not shown). **B** & **Mc** 8: 10.

Similar species. FW of eastern specimens is completely distinctive.

Range. Massachusetts to Virginia, primarily along the coast; taken inland to Fontana, North Carolina.

Status. Local; generally considered very rare, but consistently taken in some numbers at certain well-known localities (e.g., Lakehurst, New Jersey).

Season. Mid-season.

Larval foodplants. Oaks (*Quercus*), presumably Scrub Oak (*Q. ilicifolia*) in most cases. Reared on *Q. ilicifolia* (SAH, JBP).

Notes. Moth has been found resting on outbuildings in wooded areas (MCN).

C. coccinata Grote (1872)
SCARLET UNDERWING
Plate 6: 7

Description. FW light gray, mottled with whitish and brownish; with blackish shadings in areas of basal and anal dashes, and obliquely from pm line to outer margin just below apex (these shadings heavier in females). HW deep crimson with black bands. Underside with crimson ground on FW (sometimes showing through as pinkish tint in lighter areas of upperside) and lower half of HW; upper half of HW with whitish ground; all bands prominent. The name "chiquita" Bartsch was applied to an aberrational specimen with pinkish abdominal hairs. A minor variant, with a very heavy basal dash, has been named "circe" Strecker. Southern (especially Florida) specimens, with a very fine inner black band on HW, constitute the subspecies *sinuosa* Grote. 60–70 mm. **H** 34: 10. **B & Mc** 3: 16, 17 ("circe"), 18 (*sinuosa*).

Similar species. The boldly crimson HW will distinguish from any vaguely similar species (e.g., *parta*).

Range. Canada (Manitoba to Quebec, rarely Nova Scotia) to Florida (generally as *sinuosa*) and west throughout area.

Status. Uncommon to common.

Season. Early to mid-season. (Southern New England: 31 July; 6 July–4 September.)

Larval foodplants. Oaks (*Quercus*); particularly, though not exclusively, Scrub Oak (*Q. ilicifolia*). Reared on *Q. ilicifolia* (TDS).

Notes. Moth rests head-down on tree trunks.

C. miranda H. Edwards (1881)
Plate 1: 12

Description. FW uniform light gray; lines black, very fine and partly obsolete. HW black, with whitish apex and dirty whitish fringe. Underside HW blackish to base, with only whitish apex; underside FW with narrow whitish pm band (fig. 2.4B). Larger specimens from Texas have been described as a separate species, *orba* Kusnezov, but these may represent a southwestern subspecies. 40–45 mm. **H** (not shown). **B & Mc** 8: 21 (*orba*), 22.

Similar species. See *judith*. Underside (fig. 2.4D) is completely distinctive.

Range. Massachusetts (Forbes, 1954) to South Carolina.

Status. Very rare and local. Taken regularly in recent years at Fontana, North Carolina; also twice in Pennsylvania (DFS).

Season. Probably mid-season.

Larval foodplants. Unknown.

C. ultronia (Hübner) (1823)
ULTRONIA UNDERWING
DARK RED UNDERWING
PLUM TREE UNDERWING
Plate 6: 9, 11, 12, 13

Description. Very variable. FW generally gray-brown; often with darkish shading in area of basal dash, along inner margin, and in an arc setting off apex; usually with a brownish patch running along costa from pm line to apex. HW orange-red with black bands; grayish fringe, except whitish at apex. Underside with largely reddish ground; but pm band of FW and apex of HW white. Among the named forms are "celia" H. Edwards (contrastingly pale FW region between dark apex and inner margin), "lucinda" Beutenmüller (FW more evenly gray-brown, with dark shading in region of basal dash, and to a lesser extent along inner margin and beneath apex), "adriana" H. Edwards (FW entirely gray-brown, without blackish shadings), and "nigrescens" Cassino (the melanic form, with FW entirely brown-black). 50–60 mm. **H** 33: 2 ("celia"), 4 (typical, as *celia*), 7 ("lucinda," as *mopsa*). **B** & **Mc** 7: 17 ("lucinda"), 18 ("celia"), 19 ("adriana"), 20.

Similar species. Despite its variability, *ultronia* is usually easy to identify on the basis of its intermediate size, orange-red HW, and distinctive brownish FW apex.

Range. Southern Canada (Manitoba to Nova Scotia) to Florida (usually as "celia") and west throughout area.

Status. Common to abundant throughout most of range.

Season. Early to mid-season, with stragglers until quite late. (Southern New England: 10 August; 11 July–28 September.)

Larval foodplants. Various Rosaceae, including apples (*Pyrus*), and cherries and plums (*Prunus*). Reared on apple (*P. malus*) (DFS) and Pin Cherry (*P. pennsylvanica*) (TDS).

Notes. Moth rests head-down on tree trunks, usually from 6 to 12 ft. above ground, often under a large branch. Melanics ("nigrescens") seem to prefer darker backgrounds than lighter forms (Sargent, 1968).

C. crataegi Saunders (1876)
HAWTHORN UNDERWING
THORN UNDERWING
Plate 8: 1

Description. FW light greenish-gray in median area; with blackish, strongly contrasting basal area (within am line) and inner margin (to pm line); blackish shadings beyond pm line along outer margin; slight brownish shading between pm and st lines. HW yellow-orange with black bands; outer band either complete or with separate anal spot. 40–50 mm. **H** 34: 12. **B** & **Mc** 10: 5.

C. pretiosa Lintner (plate 8: 4; **B** & **Mc** 10: 4) is an intriguing moth which is often considered a form of this species, though it seems at least as close to *mira* to my eyes. Here the pale median FW area is largely whitish and extends to the inner margin, the lower portion of the dark basal area is light, and the pm-st space is largely filled with brownish. This moth was not taken by collectors for many years (from about 1920 to 1968), and was often presumed to be extinct. In 1968, however, J. Muller took two specimens in New Jersey (28 June) which correspond closely to *pretiosa* (I compared these with *pretiosa* at Yale University in 1973). Since then, Dale F. Schweitzer has taken another New Jersey specimen (6 July 1974) which clearly approaches *pretiosa* (fig. 2.5), and Charles Horton has taken a similar individual at Chapel Hill, North Carolina (12 June 1974). The status of this moth is obviously obscure. Is it a species which has remained excessively rare for a long period of time, or is it a form of *crataegi* or *mira* which has long been absent, but is now recurring? Rearings of all of the species in this complex may shed light on this matter, and are strongly recommended.

Similar species. C. crataegi may usually be distinguished from both *mira* and *blandula* on the basis of its paler median FW area (with greenish tint), and its more uniformly blackish FW shadings in the basal area and along the inner margin. The am and pm lines approach, and may touch one another at the FW inner margin in *blandula* and most *crataegi*, but are more widely separated in *mira* (and *pretiosa*). *C. mira* has a deeper orange HW than either *crataegi* or *blandula*.

Range. Southern Canada (Manitoba to Nova Scotia) to Florida and west throughout area. (Most of the *pretiosa* that I have examined have come from New York [Albany], New Hampshire, and Massachusetts.)

Status. Generally uncommon to common over most of its range.

Season. Early to mid-season. (Southern New England: 31 July; 8 July–27 August. Almost identical to *C. blandula*; averaging two weeks earlier than *C. mira*.)

Larval foodplants. Hawthorns (*Crataegus*) and apple (*Pyrus*). Reared on hawthorn (*Crataegus* spp.) (AEB, SAH, TDS) and apple (*P. malus*) (JPM).

Notes. Moth rests head-down on tree trunks.

Figure 2.5 A recent specimen (1972) of *C. crataegi* (A), an old specimen (1898) of *C.* "pretiosa" (C), and a recent specimen (1974) showing some of the characteristics of "pretiosa" (B). 1.3X.

C. mira Grote (1876)
WONDERFUL UNDERWING
Plate 8: 2

Description. FW gray, with whitish and brown shadings; latter particularly within am line and between pm and st lines; without prominent blackish shadings; light patch beyond am line along costa, extending obliquely inward to subreniform; am and pm lines clearly separated at inner margin. HW yellow-orange (deeper than *crataegi* and *blandula*) with black bands. 40–50 mm. **H** 34: 13 (as *polygama*); 35: 5 (as *jaquenetta*). **B** & **Mc** 10: 2, 3.

Similar species. Rather similar to *crataegi* and *blandula*, but FW with no strongly contrasting blackish shadings, more brown in st area, and more widely separated am and pm lines at inner margin. HW deeper orange than either *crataegi* or *blandula*. See *crataegi* (also *pretiosa*) and *blandula*.

Range. Southern Canada (Manitoba to Quebec) (absent in Maine) to Florida and west throughout area.

Status. Variable from place to place and season to season. Rare to common; apparently more common to the west in its range.

Season. Mid-season. (Southern New England: 13 August; 22 July–28 August. Averaging two weeks later than *crataegi* and *blandula*.)

Larval foodplants. Hawthorns (*Crataegus*).

Notes. Moth rests head-down on tree trunks. Taken feeding at night on wild bergamot (*Monarda*) (MCN).

C. blandula Hulst (1884)
Plate 8: 3, 6

Description. FW light gray; like *crataegi*, but without greenish tint, and with less pronounced contrast of pale median area with darker basal area and inner margin; area between pm and st lines with considerable brown shading (usually more than *crataegi*, less than *mira*); am and pm lines approach, and may touch one another at inner margin. HW and underside similar to *crataegi*. 40–50 mm. **H** (not shown). **B** & **Mc** 10: 1 (atypically brownish).

Similar species. See *crataegi* and *mira*.

Range. Southern Canada (Manitoba to Nova Scotia) to North Carolina and west throughout area.

Status. Generally uncommon to common. Apparently more common to the east in its range.

Season. Early to mid-season. (Southern New England: 1 August; 5 July–29 August. Nearly identical to *C. crataegi*; averaging two weeks earlier than *mira*.)

Larval foodplants. Hawthorns (*Crataegus*) and apple (*Pyrus*). Reared on hawthorn (*Crataegus* spp.) (SAH) and apple (*P. malus*) (JPM, TDS).

Notes. Moth rests head-down on tree trunks.

C. grynea (Cramer) (1779)
Plate 7: 22

Description. FW dull greenish-gray, rather uniform; with paler reniform-subreniform area; st diffuse whitish, sometimes nearly obsolete; am and pm lines very fine, nearly meeting at inner margin; prominent orange-brown shading along inner margin from am to st line; no basal dash. HW yellow-orange with black bands. Underside light orange and blackish, all bands prominent. An aberration with the HW almost totally black is "constans" Hulst (fig. 5.3). 40–50 mm. **H** 35: 6. **B & Mc** 9: 16; 10: 10 ("constans").

Similar species. C. praeclara is somewhat similar, but shows the following FW differences: ground lighter green; am and pm lines clearly separated at inner margin; brown shading along inner margin darker and broken between am and pm lines, but extending inward from am line to base; basal dash short but distinct. In addition, *praeclara* has a paler orange HW ground and underside. See *titania* and *alabamae*.

Range. Maine to Florida and west throughout area.

Status. Common.

Season. Early to mid-season. (Southern New England: 6 August; 12 July–8 September. Nearly identical to *praeclara*.)

Larval foodplants. Hawthorns (*Crataegus*), plums (*Prunus*) and apple (*Pyrus*).

Notes. Moth rests head-down on tree trunks.

C. praeclara Grote & Robinson (1866)
Plate 7: 23

Description. FW light greenish gray; contrastingly shaded with dark brown patches, particularly along inner margin (except between am and pm lines); short, fine basal dash. HW light yellow-orange with black bands. Underside pale whitish orange with fuscous bands. 40–50 mm. **H** (not shown). **B & Mc** 9: 32.

Similar species. See *grynea*, *titania* and *alabamae*. *C. praeclara* is the only species in this group with a distinct FW basal dash.

Range. Southern Canada (Ontario, Nova Scotia) to Florida and west throughout area.

Status. Rather local, often associated with barrens, bog edges, and other acid soil areas. Rare to common; more common to the east in its range.

Season. Early to mid-season. (Southern New England: 7 August; 13 July–13 September. Nearly identical to *grynea*.)

Larval foodplants. Reared on Shadbush (Service-berry) (*Amelanchier* spp.) and Chokeberry (*Aronia* spp.) (DFS).

Notes. Moth rests head-down on tree trunks.

C. *titania* Dodge (1900)
Plate 7: 9

Description. FW like *grynea*, but duller and with less contrast; lines obsolescent; little if any brown shading in st area and along inner margin; no basal dash. HW like *grynea*, with somewhat paler yellow-orange ground. A minor variant with somewhat more distinct FW lines is "distincta" Schwarz (this may be *alabamae*). 20–35 mm. **H** (not shown). **B & Mc** 8: 19; 9: 34.

Similar species. Usually regarded as a subspecies or variety of *alabamae*. Specimens referred to *titania* are usually smaller and plainer than *alabamae*, though intergrades occur. *C. grynea* and *praeclara* have far more prominent FW contrasts than either *titania* or *alabamae*.

Range. Mid-western states. Recorded from Pennsylvania (DFS), Illinois and Tennessee in our area.

Status. Very rare and sporadic, except at western edge of our area.

Season. Apparently early to mid-season.

Larval foodplants. Apparently hawthorns (*Crataegus*).

Notes. Moth rests head-down on tree trunks (Schwarz, 1916).

C. *alabamae* Grote (1875)
Plate 7: 10

Description. FW like *titania*, but with lines more contrasting, and some dark brown shading in st area and along inner margin (between pm and st lines); no basal dash. HW yellow-orange (paler than *grynea*, like *praeclara* and *titania*) with black bands. 30–40 mm. **H** 32: 15. **B & Mc** 9: 17.

Similar species. See *titania*.

Range. Southwestern states. Recorded from Tennessee, South Carolina and Florida in our area.

Status. Rare and sporadic in the east.

Season. Apparently early to mid-season.

Larval foodplants. Presumably hawthorns (*Crataegus*).

C. *dulciola* Grote (1881)
Plate 7: 11

Description. FW pale gray; with prominent, partially double, and evenly rounded am line; basal area somewhat darker and browner; heavy, short basal dash. HW yellow-orange with black bands. 40–45 mm. **H** (not shown). **B & Mc** 9: 31.

Similar species. The FW pattern is quite unique.

Range. Northeastern New York to New Jersey, and west and south through Ohio, Illinois and Missouri.

Status. Very rare and local.

Season. Apparently early.

Larval foodplants. Unknown.

C. *clintoni* Grote (1864)
Plate 7: 5

Description. FW pale gray, with fine black lines and dashes; basal dash long, often succeeded by a second short dash which crosses am line; blackening of veins prominent toward outer margin. HW pale yellow-orange with black bands; inner black band ending abruptly before inner margin; large apex and fringe yellowish-orange. Underside largely pale whitish-orange, with narrow blackish bands. 50–55 mm. **H** (not shown). **B & Mc** 7: 14.

Similar species. C. illecta and *abbreviatella* are vaguely similar, but neither has a FW basal dash.

Range. Southern Canada (Ontario) to Florida and west throughout area. Not known from New England.

Status. Rare to uncommon over most of range. Perhaps overlooked in some instances because of early flight season.

Season. Early (being the earliest *Catocala* on the wing; flying in June to the north).

Larval foodplants. Plums (*Prunus*) and apple (*Pyrus*).

C. *similis* W. H. Edwards (1864)
Plate 8: 20

Description. FW ash-gray, with pale gray and brown shadings, brown particularly in pm-st area; small pale triangle extending along costa at apex; reniform extending upward toward costa, giving drop-like appearance. HW yellow-orange with black bands; outer band usually with widely separated anal spot. Two named forms, which are generally restricted to the southern portions of the range, are "aholah" Strecker (with prominent blackish FW patch between reniform and pm line) and "isabella" H. Edwards (with light gray FW ground). 40-45 mm. **H** 35: 2 ("aholah"), 3 (typical, as *aholah*). **B** & **Mc** 10: 6, 7 ("aholah"), 8 ("isabella").

Similar species. Recognizable in all varieties and forms by the distinctive triangular costo-apical patch and drop-shaped reniform.

Range. Southern Canada (Ontario, Quebec), Maine (rare) to Florida and west throughout area.

Status. Generally uncommon to rare; somewhat sporadic; may fluctuate widely in abundance at any one place from year to year.

Season. Early to mid-season. (Southern New England: 31 July; 15 July-27 August.)

Larval foodplants. Oaks (*Quercus*). Found and reared on Post Oak (*Q. stellata*) (DFS).

Notes. Moth rests head-down on tree trunks.

C. *minuta* W. H. Edwards (1864)
LITTLE UNDERWING
Plate 8: 7, 10, 13, 16

Description. FW gray-brown, with prominent blackish patches or shadings; st usually contrastingly white for some distance inward from costa, forming conspicuous concave arc facing toward apex. HW dull yellow-orange, with black bands; inner band rounded at apex; outer band usually complete, without separate anal spot. Among the named forms of this variable species are "parvula" W. H. Edwards (with very broad, prominent blackish patch along inner margin of FW), "mellitula" Hulst (FW with blackish in basal area and along st, at least at costa and inner margin), "hiseri" Cassino (FW uniformly dull gray), "eureka" Schwarz (FW largely blackish between am and pm lines), and "obliterata" Schwarz (melanic, with FW largely blackish, except for more-or-less contrasting white st). 35-45 mm. **H** 32: 17. **B** & **Mc** 9: 1, 2 ("parvula"), 3 ("mellitula"), 4, 5 ("obliterata"), 6 ("hiseri").

Similar species. Some specimens may be similar to *micronympha*. Generally, *minuta* is smaller, with white st forming a more distinctly concave arc from costa, and with inner black band of HW rounded at apex (rather than sharply angled or pointed).

Range. Long Island, New York to Florida (rare) and west throughout area.

Status. Uncommon to rare through much of east, but sometimes common where foodplant has been planted.

Season. Early to mid-season.

Larval foodplants. Honey Locust (*Gleditsia triacanthos*).

C. *grisatra* Brower (1936)
Plate 8: 19

Description. FW like miniature *ulalume*; dull bluish-gray with darker markings; dark subapical shade prominent and broad, running from outer margin in downward arc to reniform; region of anal dash also darkened. HW bright yellow with black bands. 48–55 mm.

Similar species. Dark FW shadings on a rather uniform ground give a distinctive look. More specimens, and rearing studies, are needed to establish the status of this moth precisely.

Range. Described from Athens, Georgia and Florida.

Status. Presumably rare, or very local.

Season. Mid-season (types collected in mid-June).

Larval foodplants. "It will probably prove to be an oak-feeder." (Brower)

C. *connubialis* Guenée (1852)
CONNUBIAL UNDERWING
Plate 8: 5, 8, 11, 14, 17

Description. Very variable, occurring in markedly different forms. FW in typical form (*sancta* Hulst) white, with strongly contrasting black lines, blackish splotches, and prominent brown shading between pm and st lines. In form "cordelia" H. Edwards (*amasia* Smith & Abbot), FW whitish, with splotches and shading as before, but lines faint or obsolete. In form "pulverulenta" Brower, FW uniformly greenish gray, with previous pattern faintly discernable, or absent. The form "broweri" Muller is the melanic, with FW entirely deep green-black. HW yellow-orange with black bands; outer band usually sharply broken (with *straight* edge at break), with separate anal spot. 40–45 mm. **H** 35: 1 (typical, as *amasia*). **B & Mc** 9: 19 ("cordelia"), 21.

Similar species. Some specimens may be similar to certain variations of *micronympha*. The abrupt, *straight* break in the outer black band of the HW is quite consistent in *connubialis*, and clearly differs from the usual irregular or rounded break at this point in *micronympha*.

Range. Southern Canada (Nova Scotia) to Florida and west throughout area.

Status. Generally rather rare, but very local. Perhaps increasing in numbers at present to the northeast in its range (e.g., Nova Scotia and New England).

Season. Primarily early, though scattered individuals taken until quite late. (Southern New England: 2 August; 12 July–13 September.)

Larval foodplants. Oaks (*Quercus*). Reared on Red Oak (*Q. rubra*) (AEB).

Notes. Moth rests head-down on tree trunks.

C. micronympha Guenée (1852)
THE LITTLE BRIDE THE LITTLE NYMPH
TINY NYMPH UNDERWING
Plate 8: 9, 12, 15, 18, 21

Description. Extremely variable. FW grayish, often with prominently contrasting whitish or blackish markings or patches; st usually contrastingly whitish from costa, as in *minuta*, but not forming distinctly concave arc (rather, a crude *W*). HW yellow-orange with black bands; inner band usually sharply angled or pointed at apex (unlike *minuta*); outer band may be broken (or nearly broken), with separate anal spot. Among the named forms of this highly variable species are "jaquenetta" H. Edwards (FW rather uniformly green-gray), "hero" H. Edwards (FW with whitish median area; narrow at costa, broad at inner margin — unlike *connubialis*), "ouwah" Poling (oblique blackish shading or streak across FW, interrupted only at pale subreniform), "atarah" Strecker (FW ground entirely whitish, with contrasting black lines), "gisela" Meyer (black from base of FW to pm line, contrasting with pale outer margin, except for black streak below apex), and "timandra" H. Edwards (inner black band of HW very narrow). There is also a melanic form (FW entirely blackish except for a more-or-less contrasting white st line) which has not been previously named, and for which I would suggest the name "lolita." **H** 32: 21 ("gisela"); 35: 4 (as *fratercula*). **B & Mc** 9: 22, 23 ("hero"), 24 ("hero"), 25 ("jaquenetta"), 26, 27 ("atarah," as "hero"), 28, 29 ("ouwah"), 30 ("gisela").

Similar species. With such great variability, this species will always provide some identification problems. *C. minuta* and *connubialis*, both highly variable as well, will most often be confused with *micronympha*. See those species for the major points of distinction.

Range. Southern Canada (Ontario), New Hampshire to Florida and west throughout area.

Status. Generally uncommon to common. Scarcer to the north and west in its range.

Season. Early to mid-season. (Southern New England: 6 August; 5 July–8 September.)

Larval foodplants. Oaks (*Quercus*).

Notes. Moth rests head-down on tree trunks.

C. *amica* (Hübner) (1815)
Plate 8: 22, 23, 24, 25

Description. Variable, but FW usually light to medium gray, with blackish lines and shadings; often also shaded with brown. HW yellow orange, with broad outer black band only; this band ending abruptly, leaving small, widely separated anal spot. Underside HW with inner black band indicated by black streak near costa and round spot above break in outer black band. Among the named forms of this species are "nerissa" H. Edwards ("melanotica" Reiff) (FW rather blackish), "suffusa" Beutenmuller (FW heavily shaded with blackish, especially along inner margin, and extending as streaks to costa and outer margin just below apex), and "curvifascia" Brower (prominent blackish arc on FW, running from mid-point of costa to outer margin just below apex). 35–40 mm. **H** 32: 16, 19 ("curvifascia," as *lineella*), 20 (typical, as *nerissa*). **B & Mc** 8: 20 ("nerissa"); 9: 11, 12, 13 ("curvifascia"), 14, 15 ("suffusa").

Similar species. Closely similar to *jair*. That species, however, has a straighter and far less dentate pm line, and prominent brown shading between the pm and st lines. Rather intermediate specimens from New Jersey (plate 8: 27), often referred to as "Jersey *jair*," have the pm line of *jair*, but little, if any, of the brown shading between the pm and st lines. The status of these specimens will probably have to be established by rearing studies.

Range. Southern Canada (Ontario), Maine to Florida and west throughout area.

Status. Generally common to abundant.

Season. Early to late, but predominantly mid-season. (Southern New England: 23 August; 12 July–27 September.)

Larval foodplants. Oaks (*Quercus*). Reared on Black Oak (*Q. velutina*) (TDS).

Notes. Moth rests head-down on tree trunks, often on oaks; when disturbed, tends to alight higher on same tree.

C. *jair* Strecker (1897)
Plate 8: 26

Description. Similar to *amica*, but FW with straighter and less dentate pm line, and prominent brown shading between pm and st lines. FW shape somewhat broader and more blunt than most *amica*. HW and underside like *amica*. 35–40 mm. **H** (not shown). **B & Mc** 9: 10, 20.

Similar species. See *amica*.

Range. Florida. Possibly New Jersey (see discussion under *amica*).

Status. Uncommon.

Season. May–June.

Larval foodplants. Presumably oaks (*Quercus*).

A specimen of C. *maestosa* reared from a larva (*above*) and a wild-caught specimen (*below*). Specimens reared in captivity, though beautifully fresh and well-marked, may be atypically small and dark. 1.0X.

WITH THIS SURVEY of the eastern *Catocala* completed, it seems appropriate to outline some of the major biological questions that it suggests. Five such questions may be asked:

1 How have so many closely related species come to be living together?
2 How do so many species continue to coexist, without some being eliminated in competition with others?
3 How do the various species maintain their genetic integrity; i.e., what isolating mechanisms prevent hybridization among members of different species?
4 Given the extent of species diversity within the genus, what factors are responsible for the further intraspecific diversity (polymorphism) which characterizes many species?
5 Why is polymorphism so prevalent with respect to the forewings of many species, while hindwing polymorphisms are virtually unknown?

Attempts to provide tentative answers to these questions will be made throughout the remainder of this book. The matters of competition (2) and isolation (3) will be considered in some detail in chapter six, and forewing (4) and hindwing diversities (5) will be treated in chapters seven and eight. The first question, however, is best considered here. How have so many closely related species come to be living together? This question is, of course, the question of speciation. How have the many species arisen?

Evolutionary biologists ordinarily assume that the process of speciation requires some spatial separation of the members of a freely interbreeding population into isolated sub-populations. One species must be split into two or more sub-populations which are separated by geographic barriers (e.g., a river, or desert, or perhaps distance itself) to the extent that interbreeding between members of different sub-populations is prevented for a long period of time. During this period of complete separation, the different sub-populations adapt to their local conditions, thereby accumulating genetic differences which lead to reproductive difficulties (e.g., hybrid sterility) should two sub-populations come into contact again and attempt interbreeding. At that point speciation has occurred, and selection will then operate to favor the development of isolating mechanisms (e.g., seasonal, ecological, or behavioral differences) which will minimize the chance of mating attempts between members of the now different species. Speciation in the *Catocala* has presumably reached this point for the most part, as hybrid specimens are virtually unknown.

How then has geographic isolation of various sub-populations of *Catocala* species occurred? I suggest that the answer to this question may be related to a characteristic of these moths which can be inferred from recent historical records. This characteristic is one of sporadic long-range movements, correlated with periods of great abundance. A review of the literature indicates that many species of *Catocala* have sometimes been found in considerable numbers far from what we would now consider their normal ranges. Such situations were particularly prevalent around the turn of the last century (1880–1930s). For example, one may read accounts of *sappho* occurring in numbers in Illinois (Snyder, 1897*b*) and Virginia (Smyth, 1899), of *marmorata* in New York (Foulks, 1893) and Connecticut (Ely, 1908), and of *insolabilis* "by the thousands" around Chicago (Snyder, 1897*a*). In the 1920s and 1930s, species such as *lacrymosa*, *insolabilis* and *angusi* were taken in Massachusetts (A. E. Brower, personal communication, based on specimens in the Zeissig collection), as were *marmorata*, *maestosa* and *cara* "carissima" on Long Island, New York (S. A. Hessel, personal communication, specimens in his collection).

Such records seem to have been made when the abundance of a species was increasing dramatically. Indeed, most accounts of spectacular *Catocala* "swarms" include observations of species which are far outside their usual range. A recent example illustrates this well:

> The most remarkable recent account of daytime collecting of *Catocala* to come to my attention is that of the noted Mississippi lepidopterist Bryant Mather. He writes (personal communications) that on 3 and 5 July, 1970, he collected 124 specimens (15 species) of *Catocala* during the day *in the center of downtown Jackson*, Mississippi. . . . Incredibly enough, considering the locality, he took three state records (*angusi* Grote, *obscura* Stkr., and *sappho* Stkr.), plus two species of which only one specimen each had been taken in Mississippi (*ulalume* Stkr, and *dejecta* Stkr.). (Wilkinson, 1971*b*)

In addition to such instances of unusual species in *Catocala* "swarms," many collectors have noted dramatic increases over a period of several years in the abundance of previously rare or unrecorded species in their area. Sidney A. Hessel carefully documented such an occurrence with respect to *judith* and *serena* from 1965 to 1971 at his collecting site in Washington, Connecticut. A similar recent increase in abundance has been noted with respect to *connubialis* in Nova Scotia (D. C. Ferguson, personal communication). Such occurrences may represent the building up of a local population following an invasion of the area by a few wandering individuals from an "exploding" population elsewhere.

The overall pattern which emerges from these observations is one of asynchronous and highly erratic fluctuations in the abundance of various *Catocala* species, with considerable movement and range extension correlating with the peaks of abundance. Such characteristics suggest a basis for extensive speciation in the genus. Certainly the potential for establishing small, isolated breeding populations far from other members of the species is enhanced by any tendencies of individuals to travel widely during periods of great abundance.

Speciation might be hastened in some instances by the effects of chance operating on small gene pools. Thus, the few individuals (or perhaps only one fertile female) establishing an outlying population might, by chance, be different genetically from the species as a whole ("founder principle"; see Mayr, 1963, p. 211), and the outlying population itself might be small enough initially for chance to play a role in determining certain gene frequencies ("genetic drift"; see Mayr, 1963, p. 204). Both of these effects could contribute to the rapid genetic divergence of an isolated population, and the consequent incompatibility of its members with members of the parent population.

It appears that *Catocala* do tend to disperse widely during periods of great abundance, and perhaps these periods are infrequent enough to permit the occasional development of genetically distinct, outlying populations. Extinction must be the usual fate of such remote populations, but given sufficient time, some would presumably become established. Perhaps in this way, a basis for the present species diversity of the *Catocala* can be envisioned.

Let us stealthily approach the next tree. . . . What is there? Oho! my beauty! Just above the moistened patch is a great Catocala.

Holland, 1903, "Sugaring For Moths," *The Moth Book*

The Lure of Sugar III

A TIME-HONORED TECHNIQUE for collecting *Catocala* moths is known as "sugaring." This storied practice, described stirringly in Holland's "Sugaring For Moths" (1903, *The Moth Book*, pp. 146–150), is basically a matter of preparing a sweet mixture of substances that will attract the moths. The usual procedure is to paint such an ambrosial concoction on tree trunks along woodland trails at dusk. The trees are then checked by lantern at intervals during the night, when feeding moths may be captured.

The effectiveness of sweets in attracting moths was apparently first discovered by collectors in the 1830s, when these insects were noted at recently emptied sugar-casks and honey-skeps (see Allan, 1947, pp. 94–98). Shortly thereafter, sugaring as we know it today began to be practiced. As the practice spread, collectors tried to devise ever more irresistible "brews," and even today many collectors swear by their own, often secret, recipes.

An example of a "brew," and of the idiosyncracies of a collector in this regard, is provided by Allan in *"A Moth-Hunter's Gossip"* (1937):

> And what kind of sugar must you use? Brown Barbadoes, my boy. Eschew every other kind as the root of all evil where sugaring is concerned. . . . Pay no heed to what your grocer tells you: clap a killing-bottle to his head and bid him stand and deliver Brown Barbadoes in abundance. . . .
>
> Demand from your grocer the particular brand of treacle which Mr. Fowler imports from the West Indies. Use no other, or your purchase of Brown Barbadoes will have been in vain. Buy a 1-lb. tin of Mr. Fowler's treacle and allow its contents to subside into a saucepan. Add 2-lbs. Brown Barbadoes and half a pint of water, drawn straight from the cold tap. Stir well, and heat over a fire . . . till your mixture is boiling gently. Keep it boiling for four minutes by your watch; then put the saucepan aside to cool. And when it is lukewarm (not hot) . . . you must add to it one teaspoonful of *old* Jamaica rum, and stir well.

In fairness to Allan, it must be noted that in the second edition of his book (1947), he admits that this passage dates to his "palmy days."

The only invariable ingredient in sugaring mixtures seems to be sugar. This is usually brown sugar, though some use white, and honey or molasses (treacle) is often added. Many collectors dissolve the sugar in an alcoholic beverage, usually beer or wine.

Other frequently used substances include rotting fruits (particularly bananas and peaches), and fruit or flower extracts (blueberry, pear, cherry, lilac, etc.). Beyond these more-or-less standard ingredients, there is a long list of more exotic additives. Here the fancies of collectors are legion. Some will insist on adding a splash of their favorite cologne, or a few drops of the most expensive French perfume. Others, perhaps influenced by tropical butterfly collectors (see Owen, 1971, p. 189), will add the carcass of an unfortunate frog or fish, or even some excrement (sweat, urine, dung) from a mammalian source.

I am aware of no reports of *Catocala* feeding at excrement, though one species (*C. fraxini*) has been observed feeding on dead fish in Russia (Nabokov, 1947). Payne and King (1969), in a study of Lepidoptera attracted to pig carrion, observed no *Catocala* at that source, though these moths were known to be present in their study area. It would appear that the Underwings generally avoid carrion and dung, though further study is needed. (For a thorough review of Lepidoptera feeding habits, see Norris, 1936; and for a more recent discussion of feeding at carrion and dung, see Downes, 1973.)

In my own experience, a simple mixture of brown sugar and beer is very effective in attracting *Catocala*. I generally dissolve two pounds of dark brown sugar in about six ounces of stale beer, and have found that if *Catocala* are to be taken, this mixture works as well as any. Recently, I have been using a commercially available elderberry wine (12.5% alcohol, "specially sweetened") in bait traps, and have had good success. Something sweet and slightly alcoholic (thereby creating or simulating a site of fermentation) seems entirely adequate for attracting *Catocala*. This conclusion is consistent with results obtained in experimental studies of bait effectiveness with other moths (see references and review in Norris, 1936), but more experimentation is obviously needed, particularly to identify the exact stimulus, or stimuli, to which the moths are responding.

A serious problem with sugaring, and one that makes experimenting difficult, is its unpredictability. From place to place, night to night, even season to season, the same bait sometimes will, and sometimes will not, attract moths. There are great nights when hordes of eager visitors are seen at every painted tree, and there are nights when one wonders if he has concocted the perfect moth repellent. A few nights of the latter sort may discourage one for a lifetime, and I have met veteran lepidopterists who insist that sugaring simply doesn't work.

When sugaring does work, it can be a thrilling experience. By lantern light, the eyes of the living moths glow like coals, and the colors of their wings are unbelievably brilliant against the blackness of the night (fig. 3.1). Then there is always the lure of the next tree. What great and beautiful rarity must be there? Such nights are worth waiting for, and unfortunate indeed is he who doubts that they occur.

The unpredictability of sugar as a lure for *Catocala* and all other moths which may be so attracted has been discussed for more than a century in the pages of entomological journals. Opinions vary, but the consensus seems to favor the view that poor takes at sugar occur when natural attractants are flourishing. Such things as sap flows from tree wounds, naturally fermenting fruits, aphid honey-dew, and nectar-bearing flowers are undoubtedly natural sources of noctuid nourishment, and man-made baits may not be able to compete when these sources are available. An abundance of such alternate sources may account for the seasons of poor sugaring results which occur periodically in every locality.

A poor season occurred in Leverett, Massachusetts, in 1970, though this was followed by a particularly good season in 1971. In both years, *Catocala* were collected on virtually every night of the season at both lights and sugared trees in my backyard. A trail of twelve baited trees was checked on at least four occasions each night, with the final check around midnight. Various lights were operated from the house, including four 150-watt outdoor spotlights and a 15-watt fluorescent blacklight. These lights, like the baited trees, were checked periodically until midnight. Finally, a Robinson mercury vapor light-trap was operated from dusk to dawn each night. The total catch of *Catocala* and the numbers taken at each of the light sources were remarkably similar over the two years, but the number of *Catocala* at the sugared trees differed dramatically (fig. 3.2). This sort of seasonal variability, for reasons which are usually inexplicable, is commonplace in the experience of bait collectors.

Short term variations in sugaring success, such as occur from night to night within a season, may sometimes be related to variations in local meteorological and climatological conditions. Bait

Figure 3.1 *C. retecta* feeding at "sugar." Note the extended tongue.

Figure 3.2 The numbers of *Catocala* taken at various light sources and at bait during the summers of 1970 and 1971 in Leverett, Massachusetts. (The number within each bar represents the number of species taken.)

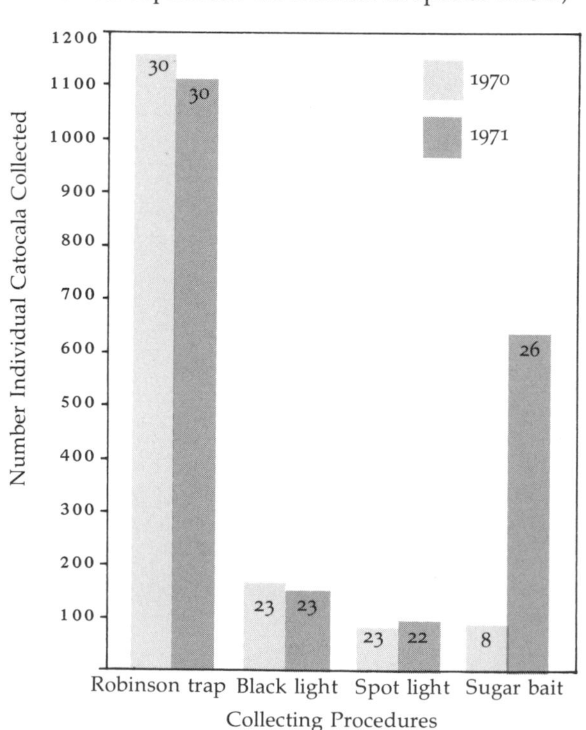

collectors generally agree that dark nights are better than bright nights, so clear weather at the time of a full moon bodes disappointment. Warm, humid nights seem to be better than cool, dry nights; still air better than wind. Light to moderate rain, or fog, is desirable, and thunderstorms in the offing seem to help. These generalizations are not without exception, and one is reminded of the oft-quoted adages as to when the fish will bite. Perhaps it is because of these uncertainties that sugaring, like fishing, is such an enjoyable sport.

One interesting question in the discussion of sugaring for *Catocala* is the extent to which it is comparable to other collecting procedures in sampling a given area for these moths. More of my own data from Leverett, Massachusetts, can be brought to bear on that question, as I have been using a number of collecting methods at that locality for several years. (These data are presented in full in Appendices One and Two, and the examples which follow are drawn from those sources.)

It is clear that bait collecting results are far less consistent from year to year than those obtained at light sources. This conclusion is supported by the data in figure 3.2, and was documented in detail by one of my former graduate students, Charles G. Kellogg (Kellogg & Sargent, 1972). At its best, bait collecting may give a better representation of individuals and species during a season than conventional light sources (incandescent bulbs and fluorescent tubes) operating at the same times of night, but often bait is far less effective. Neither bait nor conventional lights can compete, in my experience, with the overall effectiveness of a mercury vapor light-trap in sampling *Catocala*.

Certain species do seem to be particularly partial to bait, and these might be overlooked or judged to be rarer than in fact they are if collecting were confined to lights. On the other hand, there are certain species which are almost never seen at bait, perhaps in part because their flight periods occur after most bait collecting operations are terminated (i.e., after 1:00 or 2:00 A.M.).

A species by species comparison of bait and light samples for a year (1971) when both sources were particularly effective clearly reveals the kinds of discrepancies that may occur (fig. 3.3). For example, four of the five most common species at bait that year were not among the ten most common species at lights. One species, *relicta*, ranked fourth in abundance among the 26 species

Figure 3.3 The numbers of each *Catocala* species taken at light sources and bait during 1971 in Leverett, Massachusetts. The species are arranged in decreasing order of abundance at the light sources.

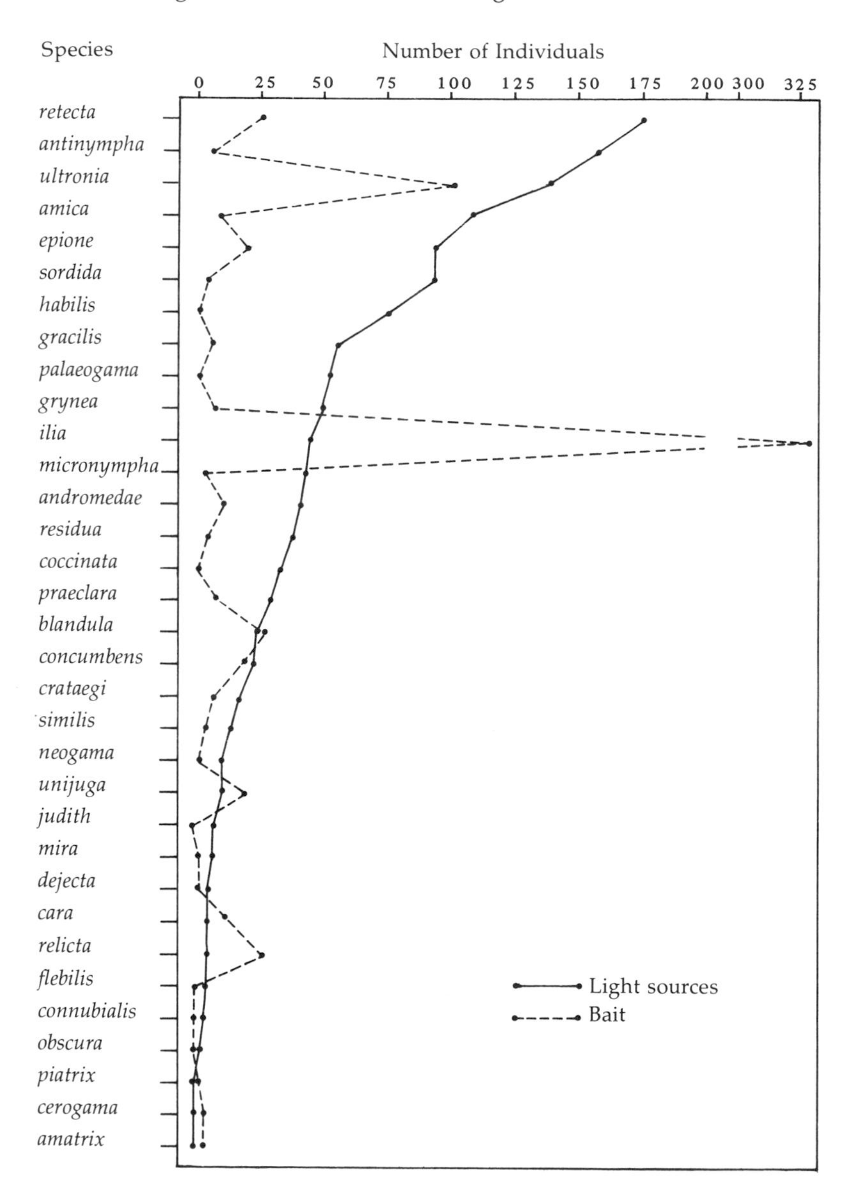

taken at bait, but ranked only twenty-seventh among the 33 species taken at lights. On the other hand, seven species were missed at bait, including two (*habilis* and *palaeogama*) which were among the ten most common at lights.

These results suggest that bait, while giving a less complete picture of species composition than several light sources, may provide some data of considerable value for assessing the abundance of certain species. Among the species which have been consistently more common at bait than at lights are *ilia*, *cerogama*, *relicta*, *unijuga*, *cara* and *amatrix*; many others, including *habilis*, *judith*, *flebilis*, *residua*, *dejecta*, *palaeogama* and *coccinata*, have been much rarer at bait than at lights. It seems obvious that one should utilize as many collecting procedures as possible when attempting to document the presence and status of the *Catocala* species in a given area. Certain procedures which have not been extensively used in Leverett, particularly daylight search and "tapping" trees for resting adults, may yield additional valuable records (see, e.g., Wilkinson, 1971*b*).

Another matter of interest regarding bait collecting concerns the sex ratios obtained by this method. The ratio of males to females in any population should be one to one, but different collecting procedures are known to yield samples which deviate from that ratio. For example, light traps usually yield a preponderance of males (Hamilton & Steiner, 1939; Williams, 1939), and collectors often turn to bait as a source of females.

I have been routinely recording the sex of *Catocala* specimens in Leverett, Massachusetts, for several years, with results that are summarized in figure 3.4 (full details are available in Appendix Two). As expected, the mercury vapor light-trap specimens were almost always male, but surprisingly, bait yielded only a slightly higher percentage of females than the fluorescent blacklight, and a significantly lower percentage of females than the incandescent spotlights. These data indicate, contrary to expectation, that bait is no more effective as a source of female *Catocala* than conventional lights operating at the same times of night.

I have no ready explanation for the different sex ratios obtained at the various lights in Leverett, but prior workers have demonstrated that different light sources may attract different proportions of the sexes in other species (Edwards, 1962). It should also be noted that the mercury vapor light-trap operated over the entire night, while the bait and the conventional lights were ordinarily operated only until midnight. It is possible that *Catocala* flying after midnight are almost exclusively male, though the magnitude of the difference in sex ratios between the light-trap and the other sources seems greater than would be expected even if this supposition were true.

Certain species of *Catocala* deserve particular attention because of peculiarities in their sex ratios. For example, though equality of the sexes characterized most species at bait, some departed from the rule and were predominantly male, e.g.:

blandula 16 males, 7 females (69.6% male)
ultronia 73 males, 21 females (77.7% male)
concumbens 11 males, 3 females (78.6% male),

or predominantly female, e.g.:

ilia 109 males, 239 females (68.7% female)
andromedae 2 males, 11 females (81.8% female)
epione 2 males, 14 females (87.5% female).

Similarly variable departures from a one to one sex ratio occurred in several species collected at the blacklight.

The significance of such variation in the sex predominating in collected samples is not immediately apparent. However, these results may indicate that behavioral differences exist between the sexes of the species involved, and these differences may be related to a separation of the species in terms of the timing of courtship and mating activities. Full understanding in this area awaits the acquisition of much more information.

The rather surprising sex ratio obtained in the sample of resting moths (figure 3.4) may be an artifact, resulting from the release of most of my Robinson trap specimens (which were almost exclusively male) into the woods around my home each morning. It is possible, however, that behavioral differences between the sexes in the selection of resting sites might account for this disproportionate sex ratio. This possibility is discussed in some detail in chapter seven.

One or two suggestions can be made regarding future efforts in bait collecting for *Catocala*. I would particularly urge collectors to check their bait during the early morning hours, and to record carefully the species and the sex of all specimens encountered at these times. Such data are sadly wanting, and there are hints that certain species may be missed by most collectors because they come to bait so late (e.g., *insolabilis* in Michigan, M. Nielsen, personal communication). It is unfortunate, in this regard, that most bait-traps, while effective for some butterflies and moths (Platt, 1969; Owen, 1971), are not particularly effective for *Catocala*. It appears that the Underwings are adept at finding their way out of traditionally designed bait-traps (D. Schweitzer, personal communication), and that experimenting with new designs would be worthwhile. Certainly an effective bait-trap would permit more controlled study of the variables affecting "sugaring" than has heretofore been possible. Perhaps this matter will challenge some of our more inventive collectors.

Figure 3.4 Percentages of males and females in *Catocala* samples taken in Leverett, Massachusetts, . . . from 1970 to 1973 at various light sources, at bait, and while resting. (The number of sexed individuals in each sample is given in parentheses beneath each collecting procedure.)

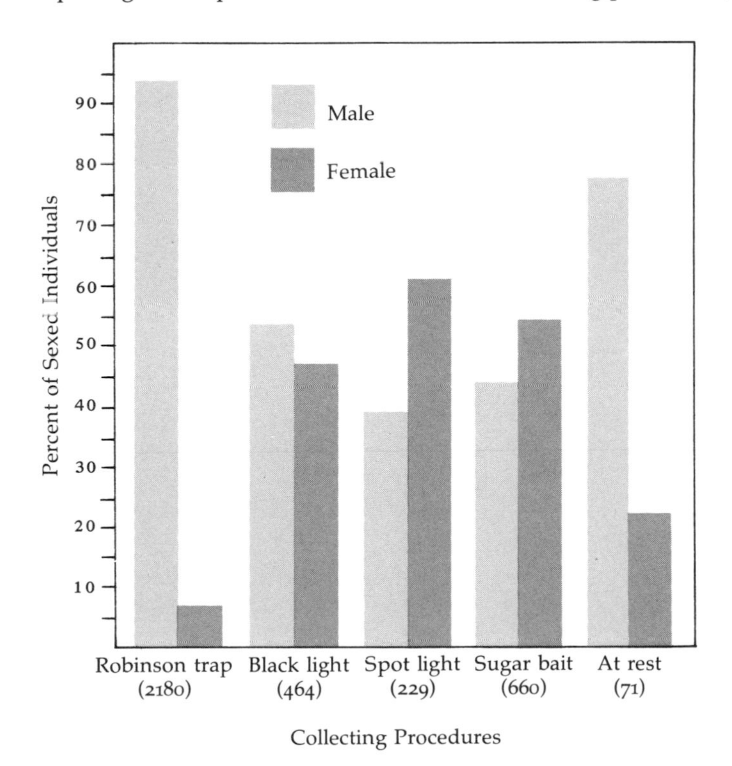

Night hath a thousand eyes.

John Lyly, c. 1589, *Maydes Metamorphose*

The Lure of Light

IV

"DRAWN LIKE A MOTH TO A FLAME" — so runs the familiar refrain based on one of man's most ancient and universal observations. Some see high drama in the fateful journey of a moth to a flame (e.g., Thomas Carlyle, "Tragedy of the Night-Moth," reproduced in Holland, 1903, p. 209), while others are more impressed with the apparent senselessness of this behavior. Among the Navaho Indians the word for moth means "the particular one which is fire crazy," and an association of moth behavior with insanity figures prominently in their mythology (Wyman & Bailey, 1964).

Entomologists have long recognized the potential of light as a collecting device, as shown by the following passage from one of the earliest entomology textbooks:

> In sultry summer nights also, if you place a candle on a table in a summer-house, or even in a common apartment, and open the window, you will often have excellent sport. . . . (Kirby & Spence, 1815, *Introduction to Entomology*)

Another early technique must have provided ample sport for more venturesome collectors:

> . . . many entomologists sally forth to the woods at night provided with a bright bull's-eye lantern, fastened in front by a leathern strap going round the waist, and armed with a clap-net to catch moths on the wing. . . . (Newman, 1841, *A Familiar Introduction to the History of Insects*)

Entomologists somehow survived this era, and their subsequent efforts to devise an ever better moth-trap have culminated in some remarkably effective designs. One of these, the Robinson mercury vapor light-trap (plate IV), has consistently yielded larger and more varied *Catocala* catches than any other collecting procedure or device with which I am familiar.

Some collectors, particularly hardy old-timers, regard the use of light-traps as rather unsporting:

> And, indeed, although lamp traps attract moths of many kinds, they have never attracted me. There is no element of sport about them, no spirit of the chase, and for me those things are essential (Allan, 1937, *A Moth-Hunter's Gossip*)

But there is no denying that light-traps are efficient. In the words of this same author,

> . . . . there is much of scientific interest, both biological and ecological, to be learnt from the use of light.

Many characteristics may contribute to the effectiveness of a light-trap in bringing in moths (see Robinson & Robinson, 1950; Haiao, 1972). The spectral range of the lamp seems particularly important, and collectors have long recognized the superiority of lamps with near ultraviolet components (about 365 millimicrons). The intensity and size of a lamp are also important, as high energy point sources are generally more effective than low energy diffuse sources. It is apparent that lights should be widely spaced, as a single, isolated lamp is usually more effective than two or more massed lamps.

The various characteristics of an effective light-trap are related to the physiological and behavioral attributes of moths, some of which have been studied in detail. For example, the effectiveness of ultraviolet is clearly related to the high sensitivity of insect visual pigments to the ultraviolet region of the electromagnetic spectrum (see Dethier, 1963; Hollingsworth et al., 1968; Robinson & Robinson, 1950). Behavioral observations have led to the use of funnels at the base of lamps, as in the Robinson trap, for moths frequently plummet when they come close to a bright light (Robinson & Robinson, 1950).

Detailed studies of the behaviors of moths in the vicinity of lights have convinced recent workers (e.g., Robinson & Robinson, 1950; Haiao, 1972) that moths are not "attracted," but rather are compelled by characteristics of their visual and locomotory systems to approach lights which they are actually seeking to avoid. In his Mach band hypothesis, Haiao (1972) suggests that moths perceive a dark region surrounding a bright light because of lateral inhibition between adjacent ommatidia in their eyes, and that they attempt to *escape* from such a light by flying toward this dark region around the bulb. This hypothesis seems to account for the common observation that moths often circle about bright lights, and come to rest in dark areas near a light source (e.g., under egg cartons or newspaper in a light-trap). For a full discussion of various ingenious experiments which have been devised to test the Mach band hypothesis, I refer the reader to Haiao's book (1972). I find his hypothesis the most compelling of those purporting to explain the apparent attraction of moths to lights.

Data presented in the previous chapter gave some indication of the great effectiveness of lights in taking *Catocala*. During two seasons (1970–1971) of intensive collecting at both bait and lights in Leverett, Massachusetts, 2790 (79.2%) of the 3522 *Catocala* specimens recorded were at lights, and 2275 of these light specimens (81.5%) were taken in a Robinson trap. The Robinson trap also gave a better representation of species over the two years, 31, as opposed to 29 at other light sources and 25 at bait. In view of this superiority in numbers of individuals and species recorded, and because of its greater consistency in captures from year to year (see chapter three), the Robinson trap seems to be the best single device with which to sample the *Catocala* populations in a given area. Certainly, whenever detailed analyses of local population parameters are contemplated, such light-trap samples should be mandatory.

The use of a Robinson trap over several years at three localities in southern New England has yielded extensive data on the *Catocala* of those localities, some of which are summarized in table 4.1 (the complete data may be found in Appendices One, Three and Four). A single light-trap, located in an open area surrounded by mixed deciduous woodlands, was operated from dusk to dawn over most of the *Catocala* season at each of these three locations. (More details regarding the trap-sites may be found in Kellogg & Sargent, 1972.) Table 4.1 reveals that the numbers of species and individuals taken at each locality have been roughly equivalent, but that there have been important differences in species composition and relative abundance.

Differences in the status of a species from place to place are usually attributed to differences in larval foodplant availability. In some instances this is rather easily demonstrated, as in the present case of *C. badia*, which has been taken only at Washington, Connecticut. This species feeds as a larva on bayberry (*Myrica cerifera*), and the plant has not been found at the locations other than Washington. More often the relationship between the abundance of an insect and its foodplant cannot be determined precisely, because foodplant abundance is so difficult to assess. However, analysis of the present *Catocala* records according to major foodplant families (fig. 4.1) suggests some differences in the abundance of various foodplants at the three localities involved. Clearly, the dominance of hickory-walnut (Juglandaceae) feeders has been greatest at Washington, while oak (Fagaceae) feeders have been most prominent at West Hatfield, and blueberry (Ericaceae) and apple-cherry (Rosaceae) feeders have been relatively most common at Leverett.

Table 4.1 The numbers of *Catocala* taken in a Robinson trap at each of three localities in southern New England, 1970–73. The species are listed in their order of abundance, and the largest sample of each species is given in boldface.

Species and Numbers of Individuals					
Leverett, Mass.		West Hatfield, Mass.		Washington, Conn.	
ultronia	348	*amica*	**714**	*palaeogama*	**886**
antinympha	**339**	*ilia*	**324**	*amica*	388
amica	279	*habilis*	**304**	*ultronia*	**382**
retecta	**278**	*retecta*	258	*concumbens*	**322**
habilis	212	*palaeogama*	250	*residua*	**298**
grynea	**184**	*antinympha*	197	*habilis*	259
palaeogama	182	*ultronia*	184	*serena**	**213**
epione	**163**	*epione*	144	*grynea*	181
sordida	**160**	*neogama*	**99**	*judith*	**166**
gracilis	**141**	*residua*	94	*mira*	**129**
andromedae	**87**	*concumbens*	76	*neogama*	98
praeclara	**67**	*grynea*	63	*retecta*	96
residua	66	*sordida*	45	*antinympha*	88
micronympha	**65**	*micronympha*	45	*obscura*	**82**
crataegi	**55**	*crataegi,*		*badia**	60
coccinata	**53**	*blandula* &	44	*epione*	59
similis	**52**	*mira*		*andromedae*	58
ilia	50	*andromedae*	42	*crataegi*	38
blandula	**46**	*amatrix*	42	*cara*	**34**
concumbens	40	*gracilis*	38	*gracilis* &	33
neogama	28	*unijuga*	**29**	*sordida*	
flebilis	**22**	*parta*	**22**	*micronympha*	32
judith	20	*coccinata*	22	*ilia*	25
dejecta	**19**	*connubialis*	**21**	*blandula*	23
unijuga	19	*judith*	20	*praeclara*	20
mira	18	*cara*	18	*coccinata*	19
connubialis	8	*cerogama*	**16**	*flebilis*	18
relicta	5	*obscura*	15	*parta*	11
cara	4	*praeclara*	10	*subnata*	**10**
obscura	3	*relicta*	**7**	*unijuga*	9
subnata	1	*flebilis*	5	*relicta*	**7**
cerogama	1	*subnata*	5	*amatrix*	5
parta	1	*similis*	5	*dejecta*	4
		dejecta	3	*briseis*	1
		*piatrix**	**2**	*innubens*	**1**
		innubens	**1**		
		briseis	**1**		
Totals	3016		3165		4055
	33 species		37 species		35 species

*Species taken at only one location.

Figure 4.1 The distribution of *Catocala* specimens taken at three localities in southern New England (table 4.1) according to larval foodplants.

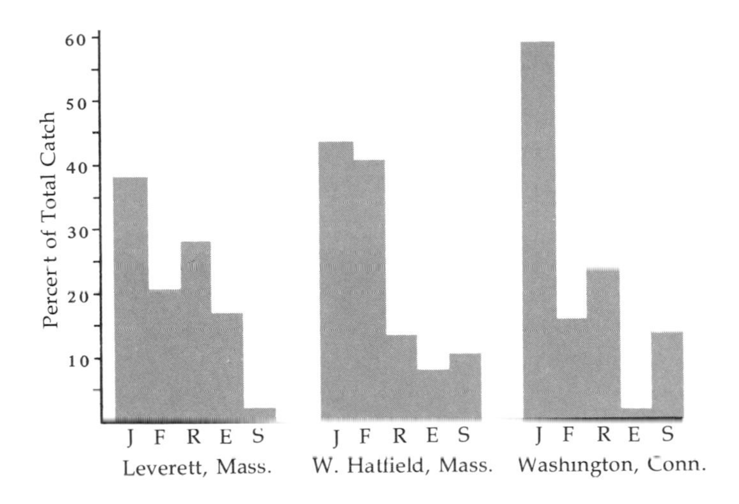

J	Juglandaceae	(Hickory, Walnut)
F	Fagaceae	(Oak)
R	Rosaceae	(Apple, Cherry, etc.)
E	Ericaceae	(Blueberry, etc.)
S	Salicaceae	(Willow, Poplar)

Differences in foodplant abundance may account for some of the differences in the relative abundance of *Catocala* species at these three localities, but other discrepancies cannot be explained in this fashion. For example, why should *serena* (a hickory-walnut feeder) be restricted to Washington, and why should *connubialis* (an oak feeder) be absent at that locality? The answers to these questions must await more detailed study of the climatological and ecological limitations on the distributions of the species involved.

The distribution of species abundance in the Robinson trap samples at each of the three localities considered here (table 4.1) is typical of most field samples of any group of animals or plants anywhere in the world, i.e., there are a few common and many uncommon or rare species. At all three locations, the 5 most common *Catocala* species comprised 50–60% of the total sample, while the 20 least common species comprised 10–20% of the total sample. This pattern was not as pronounced at Leverett, in part because there was no single, pre-eminently abundant species at that locality (as *amica* at West Hatfield, and *palaeogama* at Washington). Interestingly, 15 of the 31 species present at all three localities were most common at Leverett (as contrasted to 7 at West Hatfield and 9 at Washington), and this despite the fact that Leverett had the smallest sample of *Catocala* individuals over the period of years in question. This last result suggests that the plant communities in the sampled area in Leverett are more varied and complex, and presumably have been more stable, than those at the other locations. Such a correlation between the specific diversity of insects and the complexity (or stability) of the vegetation in an area is often assumed by ecologists. Data of the sort presented here, coupled with analyses of the age and diversity of the plant communities at each locality, would permit a direct test of that assumption. I would strongly recommend such analyses to any future students with botanical interests.

With regard to analyzing patterns of species abundance, the Robinson trap records of Sidney A. Hessel from Washington, Connecticut, taken on virtually every night of twelve *Catocala* seasons (Appendix Four), are unparalleled. From these records it is apparent that the relative abundance of a species may vary considerably from year to year at a given location. For example, *palaeogama* comprised nearly 35% of the total *Catocala* sample in

1962, but only 3% in 1968. The overall pattern of species abundance also varied over the years. The most common species in 1962 was *palaeogama*, 35% of the total sample; in 1969, the most common species, *concumbens*, comprised only 10% of the total sample. Such findings demonstrate the futility of making long-term assessments on the basis of records from a single season at any locality, no matter how extensive collecting may be during that particular season.

The frequencies of occurrence of several species for each of the twelve years during which Hessel operated a Robinson trap in Washington are shown in figure 4.2. These patterns of abundance, while quite variable in the different species, do permit some tentative generalizations. It seems fair to say, for example, that the most common species overall have been more variable in terms of annual abundance than the most uncommon species. Thus, *palaeogama*, the most common species overall, exhibits dramatic fluctuations in abundance (fig. 4.2A), and similar fluctuations characterize most of the very common species (e.g., *habilis* [fig. 4.2B], *concumbens* [fig. 4.2C], and *residua*). On the other hand, uncommon species in the overall totals maintain rather constant frequencies over the years (e.g., *retecta* (fig. 4.2E), *obscura* (fig. 4.2F), and *badia*). These uncommon species were never completely absent in any year, though such absences would have been expected if their numbers had fluctuated like those of the most common species. All of this may be saying only that species which have the potential for periodic population eruptions will be more common in the long run than species which lack (or have not yet shown) such a potential. However it is stated, the fact that the relative abundance of many species varied considerably from year to year at this location is of great ecological interest.

Annual fluctuations in numbers must interfere, of course, with attempts to detect trends of increasing or decreasing abundance over a limited period of years. The records of Hessel, as extensive as any known for the *Catocala*, are clearly inadequate for analyses in these terms. Perhaps this discovery itself is one of the most valuable findings to emerge from his records. We now know, for example, that the size of the *palaeogama* population fluctuates widely at his location, whereas less extensive records might have given an impression of great scarcity (1963–1968), great abundance (1961–1962, 1970–1971), or a dramatic trend in one or the other

Figure 4.2 The annual frequencies of occurrence of six *Catocala* species taken in a Robinson trap over twelve years at Washington, Connecticut. (The total number of each species taken is given in parentheses following the species name.)

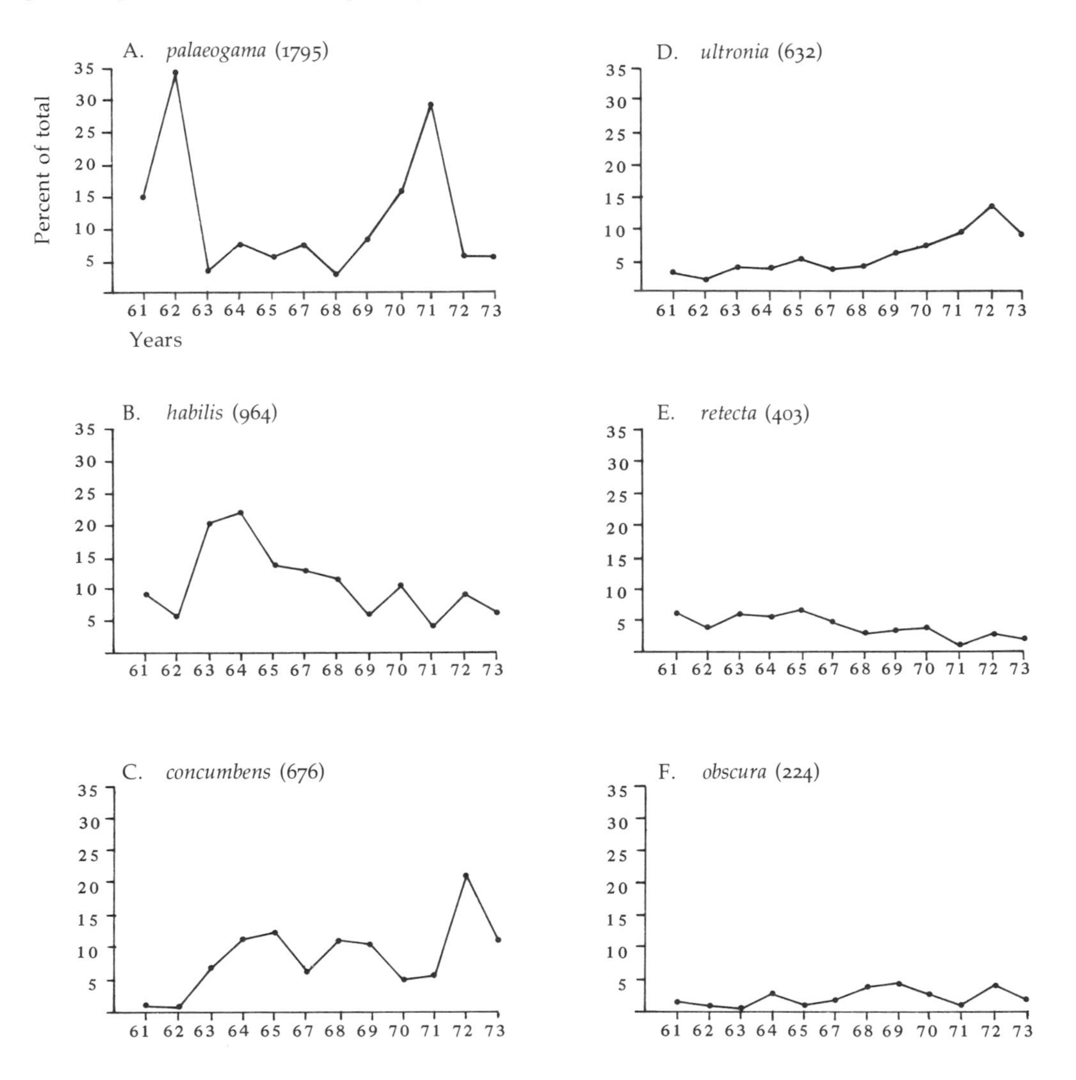

direction (1968–1971, 1963–1968). Perhaps the lesson here is to view most general assessments of status in the *Catocala* as tentative.

This chapter has demonstrated some of the potential of *Catocala* data acquired at light sources over a period of years. In concluding, I would plead with collectors to acquire more data of the sort that Hessel so patiently and carefully recorded for many years at Washington, Connecticut. These data, collected night after night and season after season, provide a historical record that can never again be obtained. In addition, they permit detailed comparisons of populations across time and space, and encourage the development and testing of hypotheses that might otherwise never arise. How much more valuable this is than acquiring specimens from hither and yon! Would that more collectors might heed the words of Thoreau:

> It is not worth the while to go round the world to count the cats in Zanzibar. (1854, *Walden*).

1
2
3
4
5

Because thou prizest things that are
Curious and unfamiliar.

Robert Herrick, ca. 1630, *Oberon's Feast*

Of Sports and Sporting V

THE CATOCALA are eminently collectable. Their size and beauty and their occurrence in myriad species, forms and varieties, have ensured their prominent place in the affections, and cabinets, of most moth enthusiasts. The genus is so large and the variety so great that there is always the lure of species not yet acquired, series not yet completed, and aberrations not yet discovered. The true collector will identify completely with P. B. M. Allan (1947, *A Moth-Hunter's Gossip*) when he writes of his long-sought *fraxini* (the white underwing of Europe, known in England, where it is excessively rare, as the Clifden Nonpareil):

> I am sure I shall find *C. fraxini* at my sugar one night, and then I shall be so scared that I shall bungle him hopelessly, and he will fly away over the tops of the trees, and I shall return home and sell my collection, and take to my bed and die. . . .

Obviously, the acquisition of specimens can become a sport to the entomologist, much as bird-listing to the ornithologist. Such sports, within reason, are generally healthy and harmless. The moth collector has an advantage over the bird-watcher, however, and therein lies a problem. For the collector may, after his own best efforts fail, turn to other means of filling his unseemly gaps — namely, purchase and exchange. These practices have been much abused, and properly maligned:

> In the passion for collection lay another evil, one that was realized to the full and reached ultimate follies. This was the danger that commercial exploitation should set in, to purvey specimens to people not remotely interested in science, but avid only of prideful and personal arrays, with emphasis, of course, upon the rare. (Peattie, 1938, *Green Laurels*)

That the *Catocala* have not escaped exploitation in this sense seems indicated by the examples of advertising from early issues of *The Lepidopterist* which are reproduced here (figs. 5.1, 5.2).

The problems with collecting are chiefly greed and obsession with the rare, and collectors have seemed peculiarly susceptible to these afflictions.

> Bear it in mind that some species appear in abundance some years, whilst in others scarcely one will be obtained, so in seasons of plenty prepare for dearth, and capture all you can; for should you get a thousand or more of a kind, it is a small number to supply the numerous entomologists in different parts of the world who may want them. (Strecker, 1878, *Butterflies and Moths of North America*)

EGGS OF CATOCALAE

Guaranteed Fertile

Prices per *dozen:* concumbens, antinympha, **20c.**; grynea, unijuga, piatrix, **30c.**; parta, amatrix, nurus, ilia, gracilis f. sordida, ultronia, cara, relicta, **40c.**; *angusi, cerogama,* relicta f. clara, ilia *var.* ex. *Arkansas,* piatrix *var.* ex. *Arkansas,* faustina, **50c.**; *diantha,* verecunda, irene, *aspasia,* **60c.**; *virgilia,* coerulea, **75c.**; *pura* **$1.00.**

Fanstina and verecunda give to some extent the var. lydia, zillah, allusa; virgilia gives *frequently* the var. valeria, volumnia.

Name of foodplants given when filling orders. *Remittance must be sent with order.*

Largest stock of fertile eggs and *perfect* **specimens of Catocalae in America. We buy, sell and exchange.**

NEW ENGLAND ENTOMOLOGICAL CO.,
366 Arborway, Jamaica Plain, Mass.

Ex-Larvae Mounted Catocalae

Perfect specimens, finely mounted.

Catocala coccinata ♂ $0.65 ♀ $0.75

Catocala coccinata var. circe ♂ $1.00 ♀ $1.25

Postage and Packing 10 cents EXTRA.

Catocalæ ova, pupæ and imagos. Bought, Sold or Exchanged
IN ANY QUANTITY

THE KATO KALA CO.,
Rudolf C. B. Bartsch, Mgr.
46 Guernsey St., Roslindale, Mass.

Figure 5.1 (*above*) An advertisement for *Catocala* eggs, appearing in the *Lepidopterist* 1, no. 2, 1916. Figure 5.2 (*below*) An advertisement for mounted *Catocala* specimens. Note the name of the company! (*Lepidopterist* 1, no. 9, 1917).

Prepare for dearth indeed!

Collectors as a class, whether of stamps and bottles, or butterflies and moths, are most covetous of that which is rare. Unfortunately, among living things the rare may quickly become the extinct, and over-zealous collecting may hasten that transition. To be sure, there are few, if any, examples of moths whose demise can be attributed solely to over-collecting, though this practice may have delivered the *coup de grâce* to some otherwise threatened populations (see Ford, 1967). At any rate, ecological awareness on the part of collectors should be encouraged in view of the potent modern threats of pollution, pesticides, and habitat destruction which so many of our species face.

An obsession with rarities poses the additional threat of blinding a collector to far more important matters regarding the common species in his area. Indeed, most collections become biased in favor of rarities, and are quite useless as sources of information on relative abundance. Peattie (1938) oversimplifies when he states, ". . . the rare is of trivial biological interest compared with the common." Nevertheless, he makes an important point which too many collectors overlook.

Misplaced emphasis often prevails with respect to aberrant individuals, commonly referred to as "sports," which appear very rarely in natural populations. To qualify as a "sport," an individual must (1) depart radically from the normal variation exhibited by its species, and (2) be so rare that its occurrence can only be the result of a freakish developmental abnormality, or a recurring mutation (fig.5.3). This last criterion is important in cases where a "sport" has a genetic basis, for any variant comprising more than 0.01% of a population, no matter how bizarre its appearance, is probably being maintained from generation to generation by direct inheritance. These variants are properly referred to as *morphs* (fig. 5.4), and must have some selective advantage which permits their retention in the population at a frequency consistent with the extent of that advantage.[1] A situation in which a species occurs in two or more distinctive morphs in a given area is referred to as *polymorphism* — a matter to be dealt with in some detail in chapters seven and eight.

[1] In actual practice, collectors are unlikely to gather samples in excess of 10,000 individuals, so decisions as to whether exceedingly rare variants are aberrations or morphs may be quite arbitrary.

Figure 5.3 The holotype specimen of *C. grynea* "constans," described by Hulst in 1884. No other specimens of *grynea* with this unusual hindwing pattern are known to exist. Such a combination of unusual appearance and great rarity makes "constans" an aberration, or "sport." 2.0X.

Figure 5.4 A specimen of *C. cerogama* "bunkeri." Despite the unusual appearance of the forewings, "bunkeri" must be regarded as a morph, as it is not sufficiently rare to qualify as an aberration or "sport." 1.3X.

"Sports," though often spectacular, are rarely of long-term significance in wild populations, as natural selection quickly operates against their survival. Their bizarre appearance usually places them at a substantial disadvantage when competing with other members of their species in such crucial matters as escaping predation and attracting mates. Perhaps on exceedingly rare occasions in the history of a species, a "sport" might have some advantage which would be selected for, and this "sport" might then become a morph. Evolution, however, normally proceeds on the basis of selection for far more subtle variations than those exhibited by "sports."

This being said, it is impossible nonetheless to deny a widespread interest in "sports." What ornithologist cannot recall the albinos he has seen? Certainly any natural monstrosity — two-headed, three-legged, four-eyed, or whatever — attracts more than passing interest. Humans being what they are, there is little doubt that "sports" will continue to thrill collectors, no matter what is said of their biological insignificance.

In the *Catocala*, the most prominent "sports" are those involving substantial alterations of the normally invariable hindwings of a species. Extreme examples would include the aberration "hilli" of *concumbens*, in which the normal pink of the hindwings is replaced with yellow; and the aberration "fletcheri" of *unijuga*, where the red of the hindwings is replaced with black. But even relatively minor alterations of a hindwing pattern, such as the slight smearing of the bands in the *concumbens* depicted on plate V (1), may be excessively rare. (Many more hindwing aberrations are described in the species accounts in chapter two.)

On the other hand, the forewings of most *Catocala* species are quite variable, and often exhibit dramatic polymorphisms. Consequently, forewing "sports" are less frequently noted, though some, such as the extreme examples of *ilia* and *agrippina* on plate V (3,4), would be difficult to overlook.

My only experience with a *Catocala* "sport" was a memorable one, and says something of "sporting" about one who should know better. On the morning of 10 September 1970, I was checking my Robinson trap, and according to my practice, I first trapped every individual in a small jar for sexing purposes. One *habilis* individual appeared ordinary when folded up in the trap, but proved extraordinary indeed when fluttering about in the

Figure 5.5 Specimens of *C. lacrymosa* (A), *C. palaeogama* (C), and a possible hybrid between these species (B). Uppersides, *left*; undersides, *right*. 0.8X.

small jar. The hindwings were almost completely black! Irregular, small patches of orange on these hindwings served to confirm my determination, but no such *habilis* had ever been described. After noting that the specimen was a male, I carefully planned his transfer to my cyanide jar. This fearsome maneuver was eventually accomplished and I waited patiently for his flutterings to cease. After what seemed eternity, the creature appeared quite dead, and I ventured to look in on my remarkable prize. Little did I realize how remarkable he was! In short, my one and only "sport" was off and away, and a subsequent day of search was all for nothing. Though *habilis* came to my trap in droves for nights thereafter, each specimen was more ordinary than the last, and so it has continued to this day. This "sport" (and I must grant him that!) has since been christened "depressans" — appropriate to the attitude it engendered upon escaping from my all-too-eager attentions. An illustration of "depressans" drawn from dictated memory, is included on plate V(2).

In addition to "sports," there are other sorts of exceedingly rare specimens which attract similar interest. *Hybrids* (i.e., offspring of matings between individuals of different species) and *sexual intergrades* (i.e., individuals in which both sexes are represented) would certainly fall into this category.

Hybrid specimens provide evidence of the breakdown of those isolating mechanisms which normally operate to prevent mating attempts across species. Isolating mechanisms are usually based on ecological (e.g., habitat) or behavioral (e.g., courtship) differences between closely related species, and these are developed and maintained by powerful selection pressures; interspecific matings are rarely fertile, and any hybrids that may be produced are almost invariably sterile.

Wild-caught hybrids are not unknown in the Lepidoptera, and in a genus such as *Catocala* where many species may occur together at one time and place, occasional hybrids might be expected. But such cases are extremely rare. I am aware of no reported cases in the literature, though I have seen some specimens which appear suspicious in this regard. Two of these specimens are depicted on the color plates (plate 3: 8, possibly *palaeogama* x *lacrymosa*; plate 4: 4, possibly *cara* x *amatrix*).

The *palaeogama-lacrymosa* specimen (fig. 5.5) was taken by M. C. Nielsen near Morenci, Lenawee Co., Michigan, on 1 September

Figure 5.6 Specimens of *C. amatrix* (A), *C. cara* (C), and a possible hybrid between these species (B). Uppersides, *left*; undersides, *right*. 0.8X.

1973. This remarkable moth has *lacrymosa*-like forewings, but the black hindwings are interrupted by narrow, broken orange bands, looking much like traces of the postmedial orange bands of *palaeogama*. The underside of this specimen is almost exactly intermediate to *palaeogama* and *lacrymosa* in both color and pattern.

Nielsen informs me that *palaeogama* is common at the site where this specimen was collected, while *lacrymosa* is very rare. This difference in the status of two closely related species would give the highest probability of hybridization, since isolating mechanisms should not be as well developed as in situations where both species are common.

On the whole, the specimen seems to have more of the characters of *lacrymosa* than of *palaeogama*, but it must be regarded as a hybrid possibility. As one person has said of the specimen, "It looks like a hybrid should!" At any rate, collectors should be on the look-out for further specimens of this sort, and efforts to mate the two species would be of considerable interest. The most likely attempt would be to mate *lacrymosa* with *palaeogama* from areas where *lacrymosa* does not occur. Perhaps *lacrymosa* could be introduced into an area where it is not now present (e.g., New England); then that area could subsequently be sampled for possible hybrid specimens.

The other suspected hybrid specimen, *cara* x *amatrix* (fig. 5.6), was taken by C. G. Kellogg at bait in West Hatfield, Hampshire Co., Massachusetts, on 23 August 1970. This specimen, clearly *amatrix*-like in many respects (e.g., forewing pattern, banding on underside of hindwing), has some of the characters of *cara* (e.g., dusky hindwing fringe and underside ground, width and shape of median black band on underside of forewing), and has impressed many observers as a very likely hybrid.

The two species, *cara* and *amatrix*, are very close, and have nearly identical daily and seasonal activity periods (Kellogg, in preparation). Certainly all specimens from localities where both species are known to occur should be scrutinized for further evidence of hybridization.

I am also aware of some hybrid possibilities between *gracilis* and *sordida*, but these two species are often confusingly similar, and at least one study (Adams & Bertoni, 1968) suggests that they may be the same species. Considerable study, particularly rearing from known parents, is needed to clarify this confusing situation.

While hybrid *Catocala* are scarce, sexual mosaics and intergrades are completely unknown. Neither *gynandromorphs* (individuals in which male and female parts develop simultaneously) nor *intersexes* (individuals which develop first as one sex and then the other) have been reported. This absence of records may be due to the very limited sexual dimorphism of most *Catocala*, a factor which would hinder the recognition of sexually intergrading specimens.

It would be difficult to overlook certain *somatic* mosaics (individuals whose wings include patches of differing genetic composition, though not of differing sexes), and these are occasionally found in the *Catocala* (e.g., plate V.5; see also Dean, 1919*b*). Such bizarre individuals are probably the most freakish of all *Catocala* "sports," and though they are insignificant in nature, they are highly prized by collectors.

In conclusion, a few comments are presented on general collecting. The acquisition of aberrations, as already made clear, may be a relatively trivial pursuit, but more thorough collecting can provide reference material of considerable scientific value. I hasten to add, however, that unless one's aim is to provide such material, he would be better advised to observe or photograph his specimens. These efforts can provide information at least as valuable as that from collecting, and they are likely to have more appeal to the increasing numbers of persons who are properly concerned with the conservation of our Lepidoptera.

If one does wish to collect, he should certainly do so correctly. Many popular works and field guides will provide information on the techniques and materials necessary to acquire, prepare, label, and house specimens. But beyond these more-or-less mechanical matters, I offer a few suggestions.

In particular, I would urge collectors to resist the temptation to discard all but immaculate specimens. If the goal is to acquire a representative series of each species in an area, then worn or damaged specimens should be retained whenever they contribute to that goal. Certain types of documentation, such as extremes of variation and seasonal occurrence, often require the retention of unsightly specimens, but these may markedly enhance the scientific value of a collection. Damaged specimens may also provide valuable information regarding predation on the insects involved (see chapter eight).

I must emphasize again the importance of obtaining quantitative data over and above the record provided by specimens. Frequencies of occurrence are never reflected accurately in a collection, and such data which may be of inestimable value to future workers should be recorded carefully. Odum (1953) has made this point: ". . . natural history becomes ecology when 'how many' as well as 'what kinds' are considered."

Finally, and at the risk of becoming trite, I would urge collectors to ponder carefully their reasons for collecting. What biological ends are the specimens to serve? What problems will they address, what questions will they answer? Otherwise, a terrible misgiving may crown long years of acquisition. Peattie (1938), in writing of the Linnaean era, expressed this matter well:

> In all this amassing of specimens, until they filled miles and miles of groaning, sagging, dusty, cluttered shelves and drawers in the museums of all Europe, there lurked a danger — that the collections should fail, at the last, to *think*. That, after all, they would represent nothing.

2
1
4
3
5

It is not clear for whom this world was made, but certainly it suits the convenience of the insects wondrous well, and to perfection they are adapted to live in it.

Peattie, 1938, *Green Laurels*

'Twixt Birth and Death VI

THE CATOCALA are moths of the sultry midsummer. Since they appear every summer, somehow they must survive the rigors of winter. Some insects solve the problem of the changing seasons simply by flying away as winter approaches, to return the following year when conditions are again propitious. One example is the Monarch butterfly, *Danaus plexippus* (see Urquhart, 1960; also Williams, 1958.) The *Catocala*, as far as we know, have no part in such evasion. Nor do they, like some butterflies and moths (e.g., the Mourning Cloak, *Nymphalis antiopa*, and the "winter moths," tribe Lithophanini), remain dormant through the season of cold and snow. How then do they appear each summer?

The answer to this question lies, of course, in the seeming magic of metamorphosis — the change of form during a lifetime. The Lepidoptera, in common with beetles, flies, bees and wasps, exhibit what is called *complete metamorphosis*, i.e., transformation through four different life-cycle stages — egg, larva, pupa and adult. This is in contrast to *incomplete metamorphosis*, the less dramatic transformation of insects such as grasshoppers, cockroaches, aphids and true bugs, whose larvae gradually develop into the adult stage. The *Catocala* go through the four distinct life-cycle stages, and it is the eggs which overwinter.

This example illustrates the basic significance of metamorphosis, which is that different life-cycle stages may have completely different functions, and may be adapted to survive in completely different situations. In the *Catocala*, as in the majority of butterflies and moths, the egg is beautifully adapted for overwintering, the larva is wonderfully equipped for feeding, and the adult is perfectly designed for dispersal and reproduction.

Catocala eggs are deposited during the summer on the trunks of trees, either singly or in small clusters, and usually tucked into crevices in the bark. These eggs are marvelously sculptured (plate VI.1), and this design, together with their small size and generally grayish color, makes them extremely difficult to detect. Crypsis such as this is no doubt important, because the eggs are subject to predation by bark-searching birds (e.g., woodpeckers, nuthatches, creepers) for many months. The eggs must also survive the physical rigors of winter, including extremes of cold and dryness. To that end, the contents of the eggs are very concentrated and so are virtually immune to freezing, and the protective coverings around each egg are almost impermeable to water (see Wigglesworth, 1972).

Figure 6.1 Two typical bark-like larvae of the *Catocala*: last instar of *C. gracilis* (*above*), and last instar of *C. amica* (*below*). 2.0X.

The larvae hatch from the eggs as the buds of their hostplants are opening in the spring (table 6.1). All *Catocala* larvae are phytophagous (plant-feeding) on broad-leafed deciduous trees and shrubs, and most species feed on representatives of one of the following plant families: Juglandaceae (hickories and walnuts), Fagaceae (oaks), Salicaceae (willows and poplars), Rosaceae (apples, cherries, etc.), and Ericaceae (blueberries). The newly hatched larvae are very tiny (generally 2 to 3 mm. long) and very active. Their activity (crawling) takes them from the site of hatching to the young leaves, upon which they immediately begin to feed. For about one month the larvae continue to feed at night and rest by day, passing through a series of molts as they grow. The period between each molt is referred to as an *instar*, and most *Catocala* go through five larval instars.

In the final instars, the larvae of most species are cryptic on twigs and bark (fig. 6.1), and possess structures, such as fringes of lateral filaments (to blur the junction of the caterpillar and its substrate) and dark saddle-marks (mimicking leaf or twig scars), which enhance this crypsis (fig. 6.2). A few atypical species (e.g., *antinympha*, fig. 6.3) rest exposed on the leaves of the foodplant. In these cases, the larvae seem to rely for protection on their ability to "jump" when disturbed, by means of powerful muscle contractions.

Molting, a process necessary to growing arthropods because of their external skeleton, may result in marked changes in appearance and behavior. In many *Catocala* (e.g., *relicta*), the first instar larva is rather translucent, showing the green color of the leaf on which it feeds, usually resting along the outer margin (fig. 6.4A). Later, as the larva becomes darker, it rests upon the mid-rib of the leaf (fig. 6.4B), and still later, it moves to the petiole (fig. 6.4C). At the time of the third molt, the larva develops a more bark-like appearance and starts to rest on small twigs. As the larva continues to grow, it rests upon larger and larger twigs (fig. 6.4D). Finally, in the last instar, the larva forsakes the twigs by day, hiding in crevices in the trunk, or leaving the tree altogether and resting in litter on the ground.

Shortly thereafter, the larva becomes quiescent, usually among dead leaves on the ground, and prepares for the next life-cycle stage, the pupa. This preparation entails some overt behavior, such as pulling together a few leaves with silk (fig. 6.4E), but the

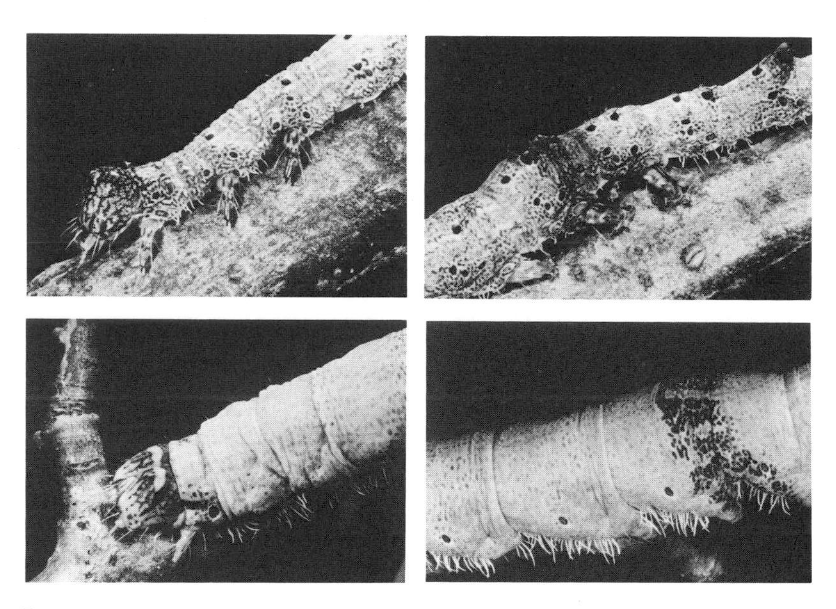

Figure 6.2 Detail of the head (*left*) and saddle-mark region (*right*) of the last instar larvae of *C. amica* (*above*) and *C. relicta* (*below*).

Figure 6.3 Third instar larva of *C. antinympha* on its foodplant, *Comptonia peregrina* (Sweet-fern). This larva, unlike most *Catocala*, is not bark-like and remains exposed on the foodplant at all times.

most dramatic events at this time are internal. When the larva molts for the last time, the new covering is the pupal shell (fig. 6.4F), and within its confines the miraculous transformation from caterpillar to moth will occur. After another three or four weeks this reorganization is complete, and the moth emerges.

The *Catocala*, as far as we know, go through one cycle of this sort each year. They are univoltine, i.e., characterized by one brood per year. While most collectors concentrate on the adults, the *Catocala* may be found in any of their life-cycle stages. Finding earlier stages usually leads to rearing the moths, which can be fascinating and often yields valuable information.

Eggs of the *Catocala* may sometimes be found by a close search of tree trunks during the fall and winter, but larvae are usually easier to obtain. One technique, much used by older collectors, is to provide hiding places for the larvae near the base of the foodplants. For example, pieces of burlap may be affixed to, or wrapped around, a tree trunk. Many larvae, when old enough to descend the tree by day, will rest in the folds of the burlap rather than travel to the leaf litter on the ground. Sometimes larvae will even pupate in the burlap, but *Catocala* pupae are otherwise difficult to find.

I have been rearing these moths for several years, and will briefly describe my techniques for the benefit of those who might wish to do the same. I obtain eggs from wild-caught females during the summer. When eggs are desired, a female is placed in a small brown bag (lunch size) out-of-doors. I often include a variety of objects with rough and irregular surfaces (e.g., cellulose sponge, bark, crumpled newspaper, etc.), as these moths often seem inclined to oviposit in crevices. If I am particularly anxious to obtain eggs from a female, I will attempt to prolong her life by feedings of a honey-water solution, provided by means of soaked pieces of sponge.

If eggs are obtained, I retain the female as a specimen (carefully labeling her so that she and her progeny can be compared), and then transfer the eggs on the small bits of paper, sponge, etc. on which they were laid, to baby food jars. These jars, covered by lids perforated with tiny holes, are kept outdoors in a somewhat sheltered location over the winter. If the jars do not receive direct sunlight and are occasionally given a little moisture, the eggs are almost certain to hatch the following spring.

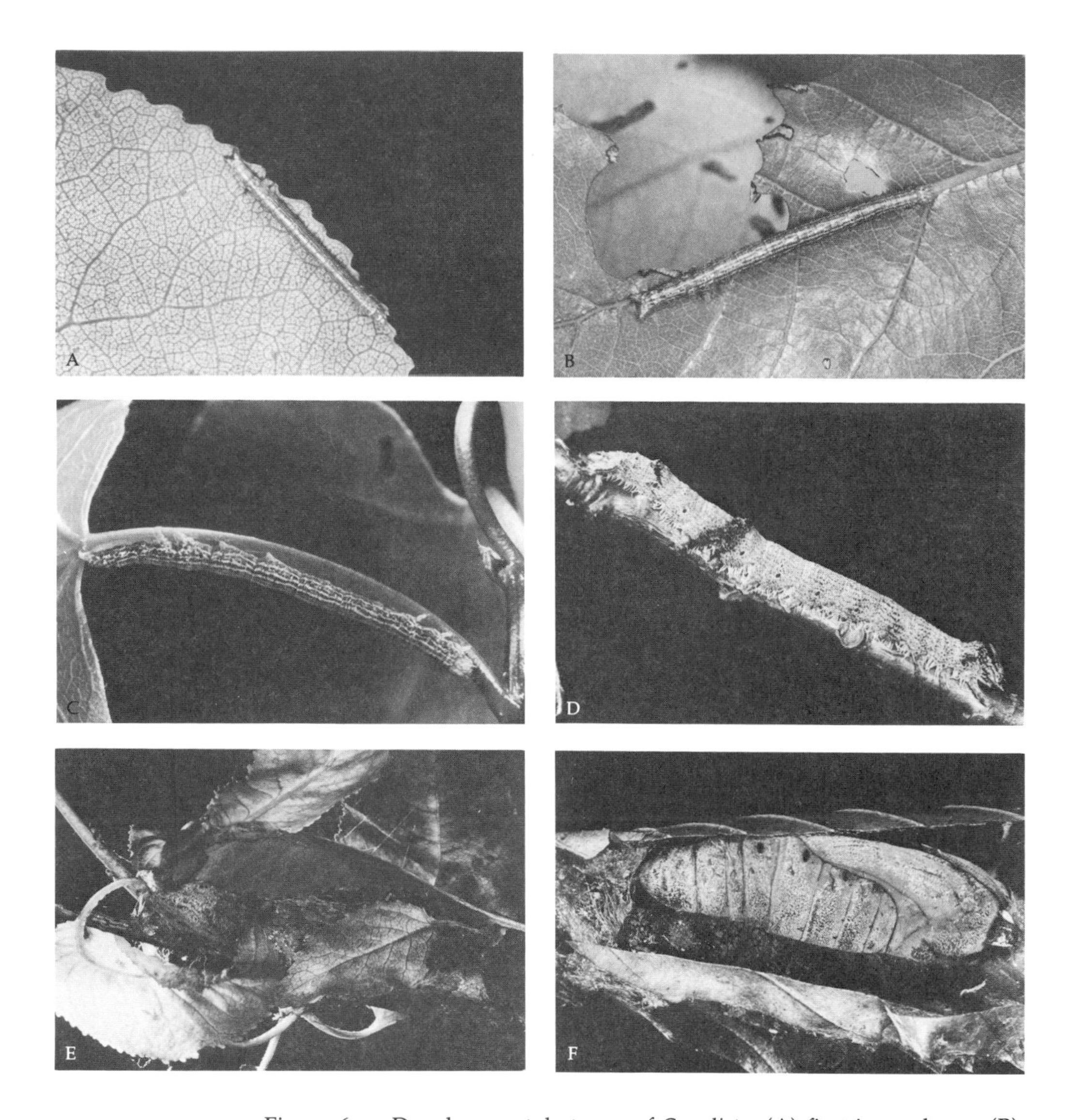

Figure 6.4 Developmental stages of *C. relicta*: (A) first instar larva; (B) second instar larva; (C) third instar larva; (D) last instar larva; (E) cocoon; (F) pupa.

When the larvae hatch, I exchange the lids with holes for solid lids, and provide fresh leaves of the appropriate foodplant. After about a week, the larvae are transferred from the jars to pint-size, plastic ice cream containers, wherein development is completed. Each plastic container houses one or two larvae, and I add twigs for resting places and crumpled newspaper or dead leaves in which they may hide and eventually pupate (fig. 6.5). Fresh food is supplied, and frass removed, on a daily basis.

These procedures have proved successful for rearing many *Catocala* species (table 6.1). The only problem I have encountered is occasional excessive humidity in the jars or plastic containers. This condition has led to the drowning of very young larvae, and to some incidence of fungal diseases in older larvae. This problem is minimized by ensuring that the containers never receive direct sunlight, and by wiping out excessive moisture when food is added each day. It may be that some species (e.g., *ilia*, *cerogama*, *amica*) would do better in screen cages, but other species (e.g., *retecta*, *relicta*, *unijuga*, *ultronia*) do very well under the conditions I have described.

Completion of the entire life-cycle of *Catocala* in captivity would require mating of the adults so that eggs might again be obtained. Mating *Catocala* in cages has proven very difficult, however. I am aware of no reports of success with any North American species (other than the case to be described here). *C. fraxini*, the White Underwing of Europe and Asia, has been mated in captivity (e.g., Cockayne et al., 1937–1938), and has even been crossed with our own White Underwing, *C. relicta* (Meyer, 1952). These reports

Figure 6.5 One of the plastic containers (cover removed) in which *Catocala* have been reared. Note the last instar larva of *C. dejecta* resting on the twig. (Photo: T. D. Sargent)

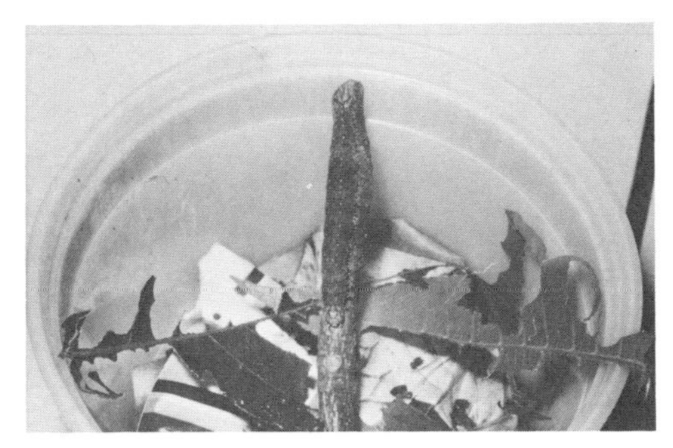

Table 6.1 Earliest dates of egg-hatching, pupation, and adult emergence in reared broods of various *Catocala* species in Leverett, Massachusetts, 1969–73.

Species (No. broods)	Earliest dates		
	Hatching	Pupation	Emergence
epione (2)	11–17 May	10 June	16–26 July
antinympha (3)	17 May	13 June	22 July
habilis (1)	23 May	6 July	4 Aug
retecta (4)	4–23 May	2–5 July	3–15 Aug
dejecta (1)	23 May	28 June	1 Aug
palaeogama (1)	18 May	20 June	31 July
ilia (14)	5–15 May	6–17 June	20–29 July
cerogama (1)*	13 May	—	—
relicta (24)	8–20 May	10–22 June	10–26 July
unijuga (2)	15 May	14 June	18 July
cara (1)*	2 June	—	—
concumbens (2)	31 May	4 July	5 Aug
amatrix (1)*	12 June	—	—
gracilis (1)	11 May	15 June	10 July
andromedae (2)	9–18 May	13 June	15–19 July
coccinata (1)	17 May	17 June	18 July
ultronia (7)	10–19 May	17–26 June	18–20 July
crataegi (1)*	11 May	—	—
blandula (1)	4 May	10 June	15 July
similis (1)*	19 May	—	—
amica (2)	17 May	18 June	22 July

*not successfully reared through larval stages

Table 6.2 Mating success with *C. relicta*, 1969–70.

Situation	No. paired	No. mated	Percent mated
Location			
Indoors	21	20	95%
Outdoors	5	4	80%
Cage			
Small	6	5	83%
Large	10	10	100%
Cylinder	10	9	90%

encouraged me to attempt matings with *relicta*, and my efforts have resulted in some success (Sargent, 1972*a*).

When mating these moths, I transferred a male and female from the individual plastic containers in which they had been reared into one of three different types of cages: *small*, homemade of aluminum window screening on a wooden frame (approx. 8 x 9 x 9 in.); *large*, obtained from Ward's Natural Science Est. (14W 7500), made of nylon mesh on a plywood frame (approx. 11 × 10 × 16 in.); and *cylindrical*, homemade from aluminum window screening rolled into a cylinder (7 in. diam. × 15 in. high) and covered at both ends with cardboard pie plates (fig. 6.6). These cages were then placed either outdoors (on stumps or hung from limbs) or indoors (basement, approx. 18°C., with small windows allowing some natural light). All paired moths were provided with the opportunity to feed from small pieces of sponge which were soaked daily with a honey-water solution.

Of 26 pairs of moths which were carefully observed during the summers of 1969 and 1970, 24 (92%) were observed in copulation, and 20 of these matings resulted in fertile eggs (i.e., produced larvae). Mating success was high in all three types of cages, both outdoors and indoors (table 6.2). Following pairing of the moths, the night during which mating occurred varied widely, and fertile eggs were obtained from matings involving males and females ranging from one to sixteen days of age. Delays in mating of two weeks or more in caged moths are very unusual (Shorey, McFarland & Gaston, 1968; Shorey, Morin & Gaston, 1968), and may account in part for the lack of success of prior mating attempts.

Courtship behaviors were not observed in detail, as a flashlight (even when covered with a sheet of red cellophane) tended to distract and disturb the moths. These behaviors, however, seemed generally similar to those described for other noctuids (e.g., Shorey, Andres & Hale, 1962; Birch, 1970). The females almost invariably adopted a "calling" posture shortly after dusk, and maintained this posture (unless mating occurred) for most of the night. In the calling posture a female elevated her abdomen above the plane of her partially spread wings, from either a horizontal or vertical surface. In high intensity calling, the wings were vibrated rapidly and the pheromone-producing gland was protruded beyond the tip of the abdomen. Just before mating, considerable male activity (walking and flying about the cage) was

Figure 6.6 The homemade cylinder cage in which *C. relicta* have been mated (*left*), and a mated pair of these moths in such a cage (*right*). (Photos: T. D. Sargent)

Figure 6.7 A *C. relicta* pair in copulation on the side of a large cage — female above, male below. (Photo: T. D. Sargent)

noted, presumably involving behaviors similar to those described in other noctuids (e.g., Birch, 1970). Copulation quickly followed, usually on the side of a cage, with the female uppermost. After initial contact the hindwings of both moths were visible beneath their partially spread forewings, but shortly thereafter the wings were closed as in the resting posture, with the female's forewings overlapping those of the male (fig. 6.7). Pairs remained in copulation for two to thirteen hours, a longer period than that reported in other noctuids (Shorey, et al., 1962; Birch, 1970).

Successful matings provide an ideal opportunity for genetic studies, as both parents of any progeny are known. I have been able to take some advantage of this opportunity with *C. relicta*. In this case, all of my reared and mated moths have been descendants of a single female taken at bait in Leverett, Massachusetts, in 1968 (fig. 6.8). All of this moth's offspring, as well as those of her progeny, have been either the "typical" or "clara" forms of this species (fig. 6.9).

The 1968 "typical" female laid 90 eggs, from which 85 progeny were reared: 42 "typical" (20 males, 22 females) and 43 "clara" (27 males, 16 females). This one-to-one phenotypic ratio suggested that a single Mendelian factor (gene) was responsible for the difference between the "typical" and "clara" forms, and that this original cross had involved a homozygote and heterozygote for that factor. This suggestion was confirmed through subsequent crosses of the progeny, the "typical" allele proving to be dominant (see Sargent, 1972, for further details).

My success with mating *C. relicta* seems to have been a case of beginner's luck, as I have had little success with other species. I have observed two matings in *antinympha*, but have failed, despite considerable effort, to obtain any matings with *epione*, *palaeogama*, *dejecta*, *unijuga*, *ultronia* and *blandula*. One reason might be that many *Catocala* species require more room for their courtship behaviors than my cages provide. Accordingly, I plan to build larger enclosures for future mating attempts. In the meantime, I hope that others will be stimulated by this challenge, and that we may ultimately develop methods of mating these moths at will. This would represent a major breakthrough, providing an opportunity to answer many outstanding questions regarding the behavior, genetics, ecology and evolution of these moths.

Figure 6.8 The wild-caught female *C. relicta* (5 August 1968, Leverett, Massachusetts) from whom several generations of reared moths are descended. 0.8X.

Figure 6.9 The "typical" and "clara" forms of *C. relicta*. These moths are a son and granddaughter, respectively, of the female depicted in figure 6.8. 0.8X.

THE FACT that so many *Catocala* species in one place are going through their life-cycle stages at approximately the same time creates two major problems for these moths. These are the problems of (1) minimizing the effects of competition for the same resources, and (2) avoiding the consequences of directing courtship and mating behaviors toward members of inappropriate species. In short, sympatric *Catocala* must develop efficient anti-competition and anti-hybridization devices. Either or both of these functions may be served by the development of various *differences* between and among species, as studies on other assemblages of insects have shown. Thus, differences in foodplants isolate the speices of the bee genus *Diadasia* (Lingsley & MacSwain, 1958); differences in the season of activity apparently separate periodical cicadas of the genus *Magicicada* (Young, 1958); similarly differences in the time of activity during the day isolate *Hyalophora* moths (Wilson & Bossert, 1963). Differences in courtship and mating behaviors are well known isolating mechanisms in many insect groups. These differences may be visual, as in *Drosophila* (Spieth, 1952); auditory, as in *Magicicada* (Alexander & Moore,1962); or chemical, as in *Hyalophora* (Schneider, 1962).

Usually, total isolation among species in nature is a result of the coaction of several different factors. In the previous examples, one factor is clearly dominant, but numerous examples could be cited where several factors coact to separate closely related species (e.g., behavioral, seasonal and altitudinal isolating mechanisms among butterflies of the *Papilio glaucus* complex; Brower, 1959).

Our knowledge regarding the factors which might separate the *Catocala* species in any one place is extremely fragmentary. Nevertheless, we will examine such data as exist, emphasizing the fact that a great deal of work remains to be done.

It seems certain that foodplant differences serve as anti-competition devices in some *Catocala*. Most of the species are limited to the plants of one family, (e.g., Fagaceae, oaks; Salicaceae, willows and poplars; etc.); at least one species is apparently isolated on a single, odd foodplant (*cerogama* on linden, *Tilia*). However, there are a number of *Catocala* that feed on the same plant species at any one location. For example, twelve species feed on Shagbark Hickory (*Carya ovata*) in Leverett, Massachusetts. Within such groups, factors other than hostplant specificity must serve to reduce competition among the species. Indeed, the existence of large numbers of species on certain foodplants, while other common trees of our eastern forests are not utilized (e.g., maples, birches, beech, etc.), suggests that competition for food resources has not been particularly intense for many of these moths.

It is possible that foodplant differences, where they occur in the *Catocala*, serve anti-hybridization as well as anti-competition functions. It has been suggested that certain male moths may respond to a combination of female pheromone and hostplant odor when seeking mates (Riddiford & Williams, 1967*a*, *b*), so that the hostplant could coact with the pheromone as an isolating mechanism. As already noted, however, substantial numbers of *Catocala* species are known to utilize the same foodplants, so these species must be isolated by other factors.

Detailed analyses of the seasonal occurrence of many *Catocala* suggest that some species may be separated in part by a seasonal offset. If complete data on adult captures are summed over several seasons, differences in the flight seasons of certain closely related species become apparent (fig. 6.10). For example, though *C. dejecta* and *C. retecta* show extensive seasonal overlap in Washington, Connecticut, it is apparent that most *dejecta* are taken long before the peak of the *retecta* season. (Three-quarters of the *dejecta* records occur before the first quarter of the *retecta* records.) Such differences, assuming that most mating is carried out early in the respective flight seasons, must help to separate some species. However, these offsets are clearly not complete, and in many closely related *Catocala* species they do not exist (fig. 6.11).

Differences in the flight seasons of certain species seem to be a result of differences in the speed of development in the larval stages. I had occasion to rear *C. dejecta* and *C. retecta* under identical conditions (including the foodplant, Shagbark Hickory, *Carya ovata*) during the summer of 1972. Data culled from my notes reveal that though both species hatched on the same date, *dejecta* developed much more rapidly than *retecta* (table 6.3).

This difference in development resulted in some interesting differences in larval resting habits. For example, when most of the *dejecta* larvae had commenced resting on twigs, most of the *retecta* larvae were still resting on the green leaves of the foodplant. Differences such as these in resting behavior should tend to re-

L	Species (No.)	July	August	September	October
W	*serena* (353)				
	habilis (964)				
W	*residua* (1217)				
	obscura (224)				
W	*dejecta* (46)				
	retecta (404)				
W	*subnata* (44)				
	neogama (446)				

5 10 15 20 25 30 4 9 14 19 24 29 3 8 13 18 23 28 3 8 13

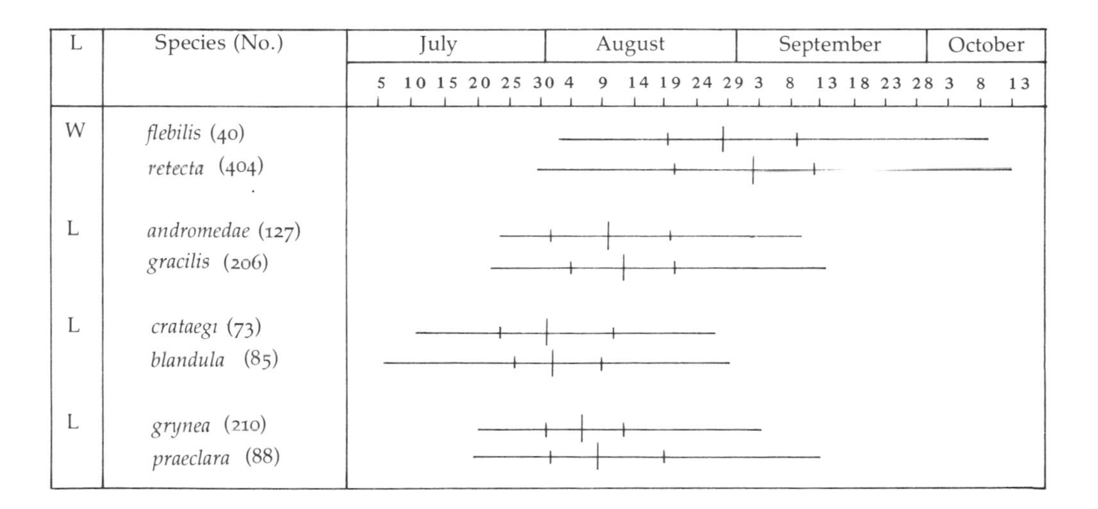

Figures 6.10 (*top*) and 6.11 (*bottom*) The seasonal occurrence of eight more closely related pairs of *Catocala* species in southern New England. The lines run from the earliest to latest dates of capture, and dashes indicate the dates by which one-quarter (short dash), one-half (heavy dash), and three-quarters (short dash) of the individuals of each species have been captured. Locations (L) are Washington, Connecticut (W) (1961–65, 1967–73), and Leverett, Massachusetts (L) (1970–1973).

duce the competitive interactions between the species; they might also serve to minimize any detrimental effect of one species on the other with respect to predation, by forcing predators to adopt two different "searching images" at any one time (see chapter 7). Ultimately, the differences in speed of development should reduce the probability of contact between adults of the two species, and so serve an anti-hybridization function.

Another possible temporal difference between species is the time of day when certain activities are carried out. The activity most frequently studied in this regard is mating, and differences in the timing of mating are known to serve as isolating mechanisms in certain closely related moths (e.g., Rau & Rau, 1929; Shorey & Gaston, 1964, 1965; Sower, Shorey & Gaston, 1970, 1971). Unfortunately, mating in the *Catocala* has been little studied, as we have seen. It is of interest, however, that the twenty-five *relicta* matings that I obtained occurred between nine and twelve o'clock at night (EDST), and that the two *antinympha* matings that I observed occurred between twelve and one o'clock in the morning (EDST).

Table 6.3 Comparison of *C. dejecta* and *C. retecta* reared under identical conditions, summer 1972, Leverett, Massachusetts.

Characteristics	*dejecta*	*retecta*
Date of first hatching	23 May	23 May
Size		
25 May	6-8 mm.	4-6 mm.
2 June	15-16 mm.	11-13 mm.
Resting site (twig/green leaf)		
6 June	22/12	0/20
8 June	24/10	1/19
11 June	27/7	12/8
Date of first pupation	28 June	5 July
Date of first emergence	1 August	15 August

There is fragmentary evidence then that the timing of mating could serve as an isolating mechanism in certain *Catocala*, but much more study is required before the significance of this possibility is fully apparent. The lack of observations and data on mating must also preclude any direct assessment of differences in chemical or visual stimuli serving as isolating mechanisms during courtship behaviors in these moths. However, certain indirect lines of evidence suggest that such factors may be important. For example, the observations that females "call," and that males display femoral tufts, or "hair-pencils" (Bailey, 1882) during courtship, suggest that pheromone-specificity may play a role in isolating various species. And the fact that hindwing colors and patterns show little intraspecific variation suggests that these structures may function as releasing stimuli, and therefore as antihybridization devices, during courtship.

It seems likely that the various factors discussed to this point, including differences in larval foodplants, daily and seasonal activity periods, and courtship and mating behaviors, are components of the complex of factors which minimize competition and prevent hybridization among the *Catocala*. There may be additional factors which have not yet been studied. Ecological differences, in the form of niche specializations of either larvae or adults, might effectively isolate some species. The use of sense modalities other than visual or chemical (e.g., auditory) might play a role in separating certain species during courtship.

This chapter emphasizes our lack of knowledge regarding the life histories of the *Catocala*. I hope this emphasis will serve as a stimulus for further studies on these insects. Collectors would materially advance our knowledge if their efforts were to include extensive rearing studies, and more detailed observations of the activities of the living moths in the field.

It is important to mention here one technique which has considerable potential for use in the field. This is the technique of color-marking the adults so that specific individuals can be recognized on later occasions. Dr. A. E. Brower did some pioneering work of this sort, and his results (1930*a*) suggest that marking techniques may yield much information regarding the longevity, ranges, and resting habits of these moths.

I color-marked a large number of *Catocala* taken at bait between 1966 and 1969 in Pelham and Leverett, Massachusetts. These moths were marked with either quick-drying paints ("Flo-Paque") applied with a brush, or waterproof inks from marking pens ("Magic-Marker"). Both substances proved easy to apply and very durable on the moths' wings. The marking was carried out to render resting moths more conspicuous (figs. 6.12, 6.13), and so more easily found on tree trunks (see Sargent & Keiper, 1969), but other data of interest were also acquired. I found that *Catocala* adults may survive for at least three weeks in nature, and that some individuals may be quite sedentary (e.g., one *cara* was taken at the same sugared tree on seven consecutive nights). On the other hand, the recapture percentage was rather low (though nowhere near as low as that acquired in light-trap collecting — S. A. Hessell, personal communication). Sixty-six of 558 marked moths (11.6%) were recaptured, and this suggests that many of

Figure 6.12 The marking of a *Catocala* with quick-drying paint (*left*), and a color-marked *C. cara* resting on Black Oak (*right*). (Photos: T. D. Sargent)

Figure 6.13 A resting *C. blandula* on white pine. This moth is color-marked but, as depicted here in black-and-white, illustrates the problem of finding resting moths which are not color-marked. (Photo: T. D. Sargent)

the moths may have wandered from the study areas. I never searched widely for these marked moths, but more ambitious persons might make better use of this technique to acquire valuable information regarding the movements of these moths.

The ease with which *Catocala* can be marked should also encourage some experimental studies of alterations in color and pattern. Such alterations would permit tests of various hypotheses regarding the functional significance of the wing markings. I have been releasing specimens of *C. relicta* with colored hindwings (e.g., plate 4: 12) into the field along with controls whose black bands are painted black, in an attempt to determine whether novel hindwings increase or decrease the chances of an individual's survival in nature. The experiment has not yet yielded sufficient data to answer this question, but the approach seems noteworthy and may stimulate others to devise similar field experiments. I sincerely hope that I do not also stimulate a spate of new species descriptions!

Oh, what a tangled web we weave,
When first we practice to deceive!

Sir Walter Scott, 1808, *Marmion*

Designs for Deceit VII

THE FINAL TWO CHAPTERS of this book will be devoted to a consideration of the anti-predator functions of *Catocala* forewings and hindwings respectively. Substantial evidence indicates that the Underwings are palatable and subject to heavy predation by birds (e.g., Jones, 1932; Sargent, 1973*b*). It seems equally clear, however, that the bark-like forewings of these moths constitute an effective first line of defense against such predation.

> Take as an illustration, the splendid moths of the great genus *Catocala*, the Afterwings, as we familiarly call them. The fore wings are so colored as to cause them, when they are quietly resting upon the trunks of trees in the daytime, to look like bits of moss, or discolored patches upon the bark. They furnish, in such positions, one of the most beautiful illustrations of protective mimicry which can be found in the whole realm of nature. (Holland, 1903, *The Moth Book*)

These marvelously cryptic forewings are undoubtedly the product of timeless encounters with sharp-eyed predators — the evolutionary product of a strategy of deceit. The problem for the predators, presumably birds, has always been one of discrimination, and just as their powers of discrimination have improved over the ages, so too have the deceptive attributes of the moths.

In particular, the moths have evolved a number of behavioral characteristics which act to maximize their cryptic resemblance to the bark of trees. One of these characteristics is the selection of appropriate resting places, particularly in terms of background reflectance. It would seem that light moths should select light backgrounds, and dark moths should select dark backgrounds. In fact, this seems to be the case, as a considerable body of observational and experimental data will attest. Our whitest Underwing, *C. relicta*, prefers to rest on the lightest trees in nature, particularly white birches in southern New England (Keiper, 1968). On the other hand, our darkest Underwing, *C. antinympha*, is found most frequently on very dark trees or blackened stumps (Sargent, 1969*a*).

Such observations, and many like them (e.g., Bunker, 1874; Cross, 1896) suggest that the background preferences of bark-like cryptic moths might be studied experimentally. Accordingly, I devised a simple apparatus (fig. 7.1) to test for background preferences. This apparatus consisted of a cylinder (19 in. high x 44 in. circumference) made up of blotting paper sections set into a plywood box (19 in. high x 15 in. square), and covered with a pane of window glass. The blotting paper sections were painted black, white, or various shades of gray, and provided a choice of backgrounds differing only in reflectance. The apparatus was set out in a wooded area where moths could be introduced into the cylinder as they were captured at night, and their background selections noted on the following morning.

In the simplest test, that is of black versus white background, dramatic results were immediately obtained (table 7.1). All light species tested, including *C. relicta*, exhibited a clear preference for the white backgrounds; all dark species tested, including *C. antinympha*, preferred the black backgrounds. (More detailed accounts of these experiments, and of variables influencing the behavior of moths in such an experimental situation, may be found in Sargent 1966, 1968, 1969*a*, and 1973*a*.)

Controversy exists as to whether background selections are based on an ability of the moths to match their own reflectance with that of a background, or on innate preferences for particular background reflectances. In one experiment, moths were painted in order to alter their reflectances (Sargent, 1968), and yet these moths continued to prefer a background appropriate to their original reflectance. This result and many others (Sargent 1969*a*, 1973*a*), suggest that the background preferences of cryptic moths are based on innate responses, though opinion on this point is not unanimous (Kettlewell, 1973). Whatever its basis, the fact that cryptic moths are able to select appropriate backgrounds can only serve to augment their crypsis.

In addition to the ability to select backgrounds appropriate to their reflectance, bark-like moths are also able to orient on a substrate so that their markings are aligned with the prevailing irregularities of the surface. This complex phenomenon has also been studied experimentally (Sargent, 1969*b*), with results which indicate that the orientation of noctuid moths on a substrate is dependent on both general gravitational cues and on direct tactile cues from the substrate itself. The result is an alignment of the pattern of the moth with the ridges and shadows of the tree trunk on which it rests. For the *Catocala*, this is effected by a head-up (rare) or head-down resting attitude, either of which serves to align the markings of moth and tree, whether these both be verti-

Table 7.1 The selection of backgrounds by light and dark bark-like species in an experimental apparatus.

Species	Backgrounds		P*
	Black	White	
Light Species			
Cosymbia pendulinaria	9	68	***
Lobophora nivigerata	0	17	***
Apatela innotata	1	19	***
Catocala relicta	0	10	**
Dark Species			
Mamestra adjuncta	9	1	*
Mamestra detracta	12	2	**
Chytonix palliatricula	27	5	***
Catocala antinympha	56	10	***

*Significant deviations from chance selections of the black and white backgrounds (chi-square tests) are indicated by asterisks for probabilities (P) of less than 0.05 (one *), 0.01 (two **), and 0.001 (three ***).

Figure 7.1 A simple experimental apparatus in which background preferences of bark-like moths have been tested.

Figure 7.2 The White Underwing (*C. relicta*) resting in its characteristic head-up position on a white birch.

Figure 7.3 An "ant's-eye" view of the *C. relicta* seen in figure 7.2. The ability to lie flat against a tree trunk is characteristic of most bark-like moths.

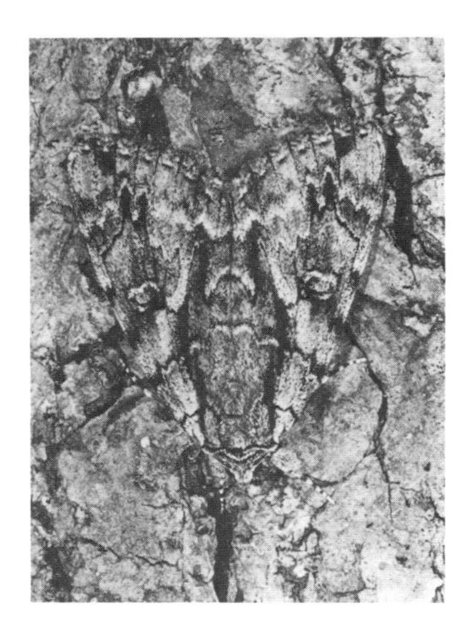

Figure 7.4 *C. retecta* resting in its characteristic head-down position on hickory. Note that the resting attitudes of the *Catocala* tend to bring their most prominent markings into alignment with those of their backgrounds.

cal (fig. 7.2) or horizontal (fig. 7.4) with respect to the moth's body axis and the trunk itself.

Behavior complements morphology, and the result is a more nearly perfect resemblance of moth and bark. Such a sophisticated defensive strategy must be both a result of, and an impetus to, increasing sophistication on the part of predators. Again, evidence suggests that moth predators, birds in particular, have evolved some highly efficient strategies with which to counter sophisticated cryptic defenses. One such strategy is known as the formation of specific searching images. This concept, though used in different ways by different workers, is basically an acknowledgement of the fact that predators often exhibit an increased tendency to take particular cryptic prey after a few encounters with that prey. It is usually, though not necessarily, assumed that this tendency is based on the development of a perceptual filtering mechanism which sensitizes the predator to items in the environment which match an image formed after a few experiences with a particular prey item. This explanation was advanced by L. Tinbergen (1960), whose field studies on the diets of Great Tits (*Parus major*) in a pine forest in the Netherlands led to the original searching image hypothesis. Tinbergen found that these birds did not take prey items in direct proportion to their actual frequencies in nature, but took less than expected of prey items at low densities, and more than expected at higher densities.

A considerable, and often controversial, literature has now developed on this subject, but the evidence for some sort of preferential responsiveness of predators towards prey items with which they have had recent experience is impressive (e.g., de Ruiter, 1952; Monk et al., 1960; Gibb, 1962; Royama, 1970; Croze, 1970; Alcock, 1973). Whatever the mechanism involved — perceptual changes (a searching image in the original sense of Tinbergen), some form of discrimination learning, or simply a preference for the familiar — the conclusion that many predators tend to be *conservative* in their selection of prey items seems inescapable. Indeed, conservatism, in this sense of selective and perseverating predation on the familiar, must be the safest and most efficient manner in which to exploit prey.

Given highly efficient predators effectively utilizing some sort of searching image behavior, is there any further counterstrategy available to cryptic prey? The answer to this question seems clearly affirmative, for conservatism of the sort we have ascribed to predators entails a vulnerability to diversity, and cryptic species are often highly variable or polymorphic with respect to their appearance and behavior.

Polymorphism is defined as the common occurrence of two or more distinctly different forms (morphs) of a species in the same area, and is exhibited, for example, by the forewings of many *Catocala* species. It appears that these polymorphisms are evolved responses to pressures exerted on the moths by predators utilizing a searching image strategy. For example, a bird might become quite expert at finding moths of one particular pattern on tree trunks, but might, at the same time, overlook individuals having distinctly different patterns. Referring to plate VII, a bird might develop a searching image for the "hero" form of *C. micronympha*, but fail to notice the typical and "gisela" forms which are also present. Experimental studies of predation in Carrion Crows (Croze, 1970) provide substantial evidence to support this interpretation of the significance of polymorphism in cryptic species.

Polymorphism, in addition to foiling predator tendencies to hunt by searching images, might also be an advantage if the conservatism of predators is based in part on rejection of the unfamiliar or novel. In that event, rare morphs of a cryptic prey species might be protected, even if discovered. Again, considerable experimental evidence tends to support this view, as numerous workers have documented a tendency of avian predators to reject unfamiliar prey (e.g., Rabinowitch, 1968; Coppinger, 1969, 1970).

From the discussion to this point, it seems possible (and perhaps desirable) to make a list of traits which would be adaptive in a cryptic species subject to attack by a predator using a hunting strategy based on searching images:

1 scarcity, and/or a highly dispersed distribution in space (to reduce encounters with the predator, and thereby reduce the probability of searching image formation);
2 variability (polymorphism) in appearance and behavior (to reduce the encounters of any one form with the predator, and so reduce the probability of searching image formation for that form; *or* to force the use of more than one searching image at a time, thereby reducing predator efficiency);

3 a short life-span, with little survival after mating and oviposition (to prevent post-reproductive individuals from contributing to the formation of a searching image which could then be used against younger, reproductive individuals);
4 as an alternative to (3), an entirely different post-reproductive appearance or behavior (to provide a searching image for post-reproductive individuals which could not be used against younger, reproductive individuals).

The *Catocala*, as a whole, seem to exhibit most of these traits. We have already noted that the majority of the species found at any one location are rather rare (chapter 4, table 4.1); certainly the most common species tend to be highly polymorphic with respect to their cryptic forewing patterns. (e.g., *ilia*, *palaeogama* and *ultronia* in southern New England). A predator, therefore, will be faced with considerable diversity, based on both inter- and intraspecific prey variability, when attempting to form searching images for these moths.

It is also clear that different *Catocala* species exhibit very different escape behaviors when disturbed while at rest. Some species invariably circle the tree upon which they are discovered and quickly alight on the opposite side (e.g., *retecta*, *habilis*, *amica*); other species take off in erratic flight and alight again only when some considerable distance away (e.g., *lacrymosa*, *amatrix*, *parta*); still other species simply drop to the ground (e.g., *epione*, *concumbens*). Behavioral variation of this sort, which creates a condition of unpredictability for a predator, has been referred to as *protean behavior* (Chance & Russell, 1939). Protean behavior is defined as "that behavior which is sufficiently unsystematic to prevent a reactor predicting in detail the position or actions of the actor" (Humphries & Driver, 1970). The term is usually applied to forms of intraspecific behavioral variation, but the concept may easily be extended to apply to multi-species assemblages, such as the *Catocala*, where unpredictability is the product of the different behaviors of several closely similar species.

A stimulus which is unpredictable (protean), unexpected (anomalous), or unfamiliar (novel) seems to be a powerful deterrent to effective predation. This effectiveness of unusual stimuli is apparently related to neurophysiological events which are elicited in the predator (arousal, conflict, etc.), and which block or otherwise interfere with normal predatory responses (Sokolov, 1960;

Coppinger, 1970; Driver & Humphries, 1970). Clearly, such a phenomenon will tend to promote diversity among prey, since the more stimuli with which a predator must contend, the more likely is it that any one of them will be unpredicted, unexpected, or unfamiliar. This tendency of predators to promote diversity in their prey has been described as *apostatic selection* (Clarke, 1962), and may explain a great deal of the extraordinary specific and intraspecific diversity of the *Catocala*. This matter will be discussed in more detail in the next chapter.

There is some evidence to suggest that the behaviors of Underwing moths change with age. Brower (1930), in a mark-release-recapture study, found that older individuals tend to disperse much more widely than younger individuals. This dispersal of older individuals might remove them from the immediate vicinity of younger individuals, and so reduce the probability of predators forming searching images for the species in areas where reproductive activities are being carried out. Dispersal might also carry the older individuals into atypical habitats, so that even if predators were to find them in sufficient numbers to form a searching image, that searching image could not be used for the detection of younger individuals in the species-typical habitat.

Some species also seem to show a change in appearance with age. This is presumably a result of wear, which causes removal of some scales from the wings, and results in a rather different overall aspect. I have noted this particularly in *C. ilia*, when older individuals have large whitish patches on the forewings which give them a distinctive appearance on tree trunks. Again, a searching image for these worn individuals might not be useful for detecting younger individuals.

The fact that collectors tend to find an overwhelming predominance of males when searching tree trunks for *Catocala* (e.g., French, 1886; also see chapter 3, fig. 3.3) may have some significance with respect to the formation of searching images by predators. Males are generally regarded as the more expendable sex, as their essential reproductive contribution is completed with the act of mating; females must carry out other behaviors, such as oviposition, for a considerable period of time following mating. It seems possible that males, particularly those which have already mated, might select different (presumably more obvious) resting sites than females. In this way, predators (including human collectors!) might form searching images for the males but overlook the differently (and more effectively) hidden females. Such a vulnerability of males, serving to protect the females, would be akin to that postulated for the brightly plumaged males of sexually dimorphic birds (see Etkin, 1964, pp. 112–13).

In summary, it is apparent that the *Catocala* moths are among the most highly sophisticated practitioners of the art of cryptic defense to be found in the animal kingdom. Building on a resemblance of their forewings to the bark of trees, these moths have evolved a series of elegantly co-adapted behaviors which serve to enhance this crypsis and to challenge the very limits of their predators' perceptual abilities. In addition, these moths have developed a measure of diversity and unpredictability in their appearance and behavior which counters their predators' strongest asset, the ability to remember systematic and repeatable relationships — in short, to learn. As if this were not sufficient defense, the hindwings serve additional anti-predator functions which we have yet to consider!

BEFORE TURNING our attention to the hindwings, it seems appropriate to consider one other matter relating to forewing variability in the *Catocala*. This is the phenomenon known as *melanism*, i.e., a darkening of the forewings by the deposition of melanin pigment which renders these wings almost entirely black. Melanism is rather frequently encountered in the *Catocala* and is part of the recent dramatic increase in the incidence of melanic individuals in populations of bark-like cryptic moths in many industrialized parts of the world.

As evidence that melanic *Catocala* have only recently become common, we may note that none were figured among the 94 *Catocala* specimens depicted in Holland's *Moth Book* (1903), and only 4 were figured among the 217 specimens illustrated in Barnes & McDunnough's treatise on these moths (1918). On the other hand, 11 of the 126 specimens depicted in the color plates of the present book are melanic individuals. A glance at the dates of the original descriptions of melanic forms in the *Catocala* (table 7.2) will also substantiate their recent appearance in numbers sufficient to be noticed by collectors.

Melanism in moths has been most extensively studied by H. B.

Table 7.2 The described melanic forms of the eastern *Catocala,* with author and year of original description.

Species	Melanic	Author	Year
palaeogama	"denussa"	Ehrman	1893
neogama	"mildredae"	Franclemont	1938
ilia	"satanas"	Reiff	1920
cerogama	"ruperti"	Franclemont	1938
unijuga	"agatha"	Beutenmüller	1907
parta	"forbesi"	Franclemont	1938
meskei	"krombeini"	Franclemont	1938
amatrix	"hesseli"	Sargent	1976
sordida	"engelhardti"	Lemmer	1937
gracilis	"hesperis"	Sargent	1975
ultronia	"nigrescens"	Cassino	1916
minuta	"obliterata"	Schwarz	1919
micronympha	"lolita"	Sargent	1976
connubialis	"broweri"	Muller	1960

D. Kettlewell and his associates in England, working particularly with the Pepper-and-Salt Moth, *Biston betularia* L. (Geometridae). From these studies, the following major findings have emerged: (1) melanism in most cases has a simple genetic basis, being controlled by a single gene, with the allele for black ordinarily dominant to that for pale or typical coloration; (2) melanic individuals appear to be more viable than typicals (Ford, 1937, 1964; Kettlewell, 1958), and may exhibit other differences — e.g., in larval habits (Kettlewell, 1958, 1961), courtship and mating behaviors (Kettlewell, 1957), and background selections (Kettlewell, 1955*a*); (3) melanic forms have increased in relative abundance in recent times, particularly in industrial areas (Kettlewell, 1958*b*); and (4) melanic individuals are at a selective advantage in industrial areas and at a selective disadvantage in rural areas, when compared with typical individuals; differential predation by birds on the two forms contributes significantly to the selective advantage or disadvantage observed (Kettlewell, 1955*b*, 1956).

A general explanation for industrial melanism, emphasizing the cryptic advantage of melanic individuals in industrial areas (Kettlewell, 1958), has now gained wide acceptance. In this view, melanics are favored wherever air pollutants darken the tree trunks by killing the epiphytic lichens and by depositing soot. Considerable data tend to support this explanation in the case of *B. betularia* (for extended reviews of this topic, see Ford, 1964, and Kettlewell, 1973), but there are a number of findings which suggest that this explanation may not suffice for all recent cases of increased melanism.

Before considering this contrary evidence, it should be noted that the term melanism is correctly applied in cases like that of *B. betularia*, where a distinctly different, totally black morph occurs, which is known (or plausibly presumed) to have a simple genetic basis (often one gene) (fig. 7.5). However, the term has also been applied in cases where an entire population becomes somewhat darker over a long period of time, presumably because of the selective advantage of slightly darker individuals over a series of generations. This latter phenomenon, beautifully demonstrated by *Catocala* taken in the vicinity of Pittsburgh, Pennsylvania, over the past fifty to seventy-five years (fig. 7.6; many specimens in the Carnegie Museum), is almost certainly related to the cryptic advantage of darker individuals in a darkening environment; it is

Figure 7.5 A typical specimen of C. *cerogama* (*above*), and its melanic form, "ruperti" (*below*). This is an example of discontinuous variation (polymorphism) in a species, as intergrades between these phenotypes are virtually unknown. 1.0X.

Figure 7.6 Recent specimens of *C. relicta* "phrynia" from Leverett, Massachusetts (*above*) and Pittsburgh, Pennsylvania (*below*). This is an example of continuous variation in a species, as intergrades between these phenotypes are commonly found. 1.0X.

also clearly different from the one-step production of melanic individuals in a melanic polymorphism. I would suggest that the term melanism be restricted to cases where a clear-cut polymorphism is involved, and that some other phrase, perhaps "progressive darkening," be used to describe the more gradual change of an entire population towards a darker appearance.

One finding that poses a problem for a general explanation of recent melanism based on cryptic advantage is the extent of melanism in ostensibly rural areas (i.e., where the trees are not devoid of lichens and are not noticeably darkened by soot). My study area in central Massachusetts is an example, and here melanic frequencies are quite high in several species, including some *Catocala* (table 7.3) (Sargent, 1971, 1974). Many of these melanics are extremely dark, often nearly jet black, and would seem to be cryptic on only the blackest trees in heavily polluted areas. Furthermore, many of them, like their typical counterparts, prefer light backgrounds in experimental tests (Sargent, 1968, 1969*d*), making it even less likely that their occurrence is explicable in terms of a cryptic advantage.

A number of other findings create further difficulty in devising a general explanation of melanism. For example, while most cases of recent melanism involve bark-like cryptic insects, at least one warningly colored beetle, *Adalia bipunctata* L., is displaying an

Table 7.3. Frequency of occurrence of the melanic forms of various species in Leverett, Massachusetts. (See Sargent, 1974, for further details.)

Species	Years sampled	Total sample	Number melanic	Percent melanic
Nacophora quernaria	1971–74	92	53	57.6
Biston cognataria	1971–74	184	8	4.3
Phigalia titea	1968–74	1127	227	20.1
Panthea furcilla	1970–73	389	242	62.2
Catocala ultronia	1968–74	586	100	17.1
Catocala connubialis	1970–74	15*	3	20.0

*Of this total, 9 were the "semi-melanic" form "pulverulenta."

increase in melanic frequencies in industrial areas (Creed, 1966). Another difficulty is posed by the discovery of a steep cline in the melanic frequencies of *Gonodontis bidentata* Clerck (Geometridae) across the city of Liverpool in England (Bishop & Harper, 1970). Finally, recent analyses of a variety of factors associated with industrialization have cast doubt on the completeness of a cryptic advantage explanation for even *B. betularia* (Bishop, 1972; Lees, Creed & Duckett, 1973).

These last few studies suggest that the incidence of melanism in some cases may be related to effects of industrialization other than observable environmental darkening. Possible pollution effects on the insects, perhaps acting on the larvae through chemical contamination of the vegetation, seem to warrant further investigation. It may be that the superior viability of melanics, stressed some years ago by Ford (1937, 1940), has been generally underestimated in recent discussions regarding the selective advantage of melanism. Older ideas attributing melanic success in industrial areas to relaxed selection pressure from predators might also bear some re-examination in light of these recent findings.

The relative dearth of information on the incidence of melanism in North American moths is unfortunate. Since the reviews of Owen (1961, 1962) called attention to the phenomenon, little else has been published. Owen & Adams (1963) analyzed the occurrence of melanism in *Catocala ilia* in Michigan, and Klots (1964, 1966, 1968*a*,*b*) documented increases in the frequencies of the melanic forms of *Charadra deridens* and *Panthea furcilla* (Noctuidae) in Connecticut. I have provided more recent data on melanism in several species in central Massachusetts (Sargent, 1974). Much more information is needed, however.

Perhaps collectors can be persuaded to keep routine counts of the various morphs of some of our most obviously interesting species (e.g., *Biston cognataria*, the North American counterpart of *B. betularia*). Certainly collectors of *Catocala* have opportunities to contribute valuable data, since many of these moths have distinctive melanic forms. At any rate, it seems quite clear that our understanding of melanism is far from complete, and we can hope that the opportunity will not be lost to study the phenomenon as it unfolds.

Presume not that I am the thing I was.

Shakespeare, ca. 1597, *2 Henry IV*

Designs for Display VIII

WEBSTER'S INTERNATIONAL DICTIONARY gives the following derivation of the word *Catocala*:

Catocala [NL., fr. Gr. *kato* below, taken in the sense of behind + *kalos* beautiful.]

"Beautiful behinds" — perhaps I missed the best title for this book! At any rate, the hindwings are indeed the most striking and unusual feature of these moths, and account, of course, for the common name Underwings. Despite the striking beauty of these structures, however, there has been little study of their functional significance.

It is generally assumed that these boldly patterned and colorful hindwings are examples of "flash coloration," revealed when crypsis fails to prevent attack and a moth takes flight, and concealed again when the moth alights. Such "flash and cover" sequences are presumed to confuse predators as to the whereabouts of the moths (Cott, 1940; Ford, 1967). It has also been pointed out that the combination of a startling flash pattern and erratic escape flight creates a highly unpredictable (or protean) situation for predators, and that such unpredictability is known to interfere with efficient predation (e.g., Humphries & Driver, 1970).

Some workers have emphasized the actual flash of the hindwings as *the* important anti-predator device, functioning to startle predators and to affect adversely the efficiency of their attacks (e.g., Coppinger, 1970). In this event, *Catocala* hindwings would have a function similar to that suggested for the large eye-spots of saturniid moths (Blest, 1957). It has also been suggested that the interspecific diversity of the hindwing colors and patterns of the *Catocala* might increase the startle function for these structures by introducing the potential of novelty (unfamiliar stimuli) or anomaly (unexpected stimuli) into the overall predator-prey system involving birds and these moths (Sargent, 1969*c*). For example, a bird having several successive encounters with moths possessing a particular hindwing pattern (e.g., yellow and black bands) might learn to overcome its startle response to that pattern (i.e., become habituated); this same bird, upon encountering a distinctly different hindwing pattern (e.g., pink and black bands) might again be effectively startled.

Although most suggestions about the function of *Catocala* hindwings have stressed potential startle effects, it is also possible that these structures serve a deflective function. In that case, predators would direct their attacks toward these prominent structures, thereby being directed away from more vulnerable body parts. *Catocala* hindwings would then be functionally related to the colorful, but expendable, tails of certain lizards (Cott, 1940).

It appears likely that *Catocala* hindwings serve one or more of the preceding anti-predator functions, but it is also quite possible that they play an important role in other aspects of the biology of these moths. The rather surprising lack of intraspecific hindwing variation in the genus suggests that these structures may function in courtship and mating behaviors. Little is known of such behaviors, although at least one species. *C. relicta* Walker, is able to mate under conditions of complete darkness (Sargent, 1972). Until much more is known of *Catocala* courtship and mating behaviors, the question of a sexual role for the hindwings must remain open.

One of the chief problems which has plagued efforts to assess the functions of the *Catocala* hindwings is the difficulty of directly observing the use of these structures in sexual or anti-predator contexts. We have already noted the lack of observations of mating behaviors in these moths (chapter 6), and recorded instances of predation on the *Catocala* are extremely rare. However, it occurred to me some years ago that an indirect approach to determining some aspects of hindwing function might be provided by the beak-damaged specimens which are not infrequently encountered in large samples of these moths (Sargent, 1973*b*).

Beak-damaged Lepidoptera, characterized by crisp tears and imprints on otherwise fresh wings, have always attracted considerable attention, though most studies have been concerned with butterfly examples. In many instances, interpretations of various damage patterns have been advanced, and these have, on occasion, provided some insight into the anti-predator functions of certain wing markings. For example, the incidence of beak-tears involving the small, eye-spot markings on the margins of some butterfly wings (fig. 8.1) have suggested that these markings direct the attacks of birds (e.g., Marshall & Poulton, 1902; Swynnerton, 1926; Blest, 1957), and so function as deflective devices. On the other hand, crisp beak-imprints on the wings have been interpreted as evidence that their bearers are unpalatable, since such specimens appear to have been purposely released by birds (e.g., Collonette & Talbot, 1928). This explanation receives support from studies showing a relatively high frequency of crisp beak-imprints on the wings of warningly colored (unpalatable) butterflies (e.g., Carpenter, 1941). Although beak-damage patterns in moths have received little attention, these studies on butterflies suggested that some insight into the predator-prey relationships of the *Catocala* moths might be gained through an analysis of beak-damaged specimens.

Accordingly, I decided to retain every *Catocala* specimen exhibiting suspected beak-damage during my collecting efforts over the summer of 1971. At the same time, I experimented by releasing *Catocala* moths to Blue Jays (*Cyanocitta cristata*) being housed in two large aviaries (each 8 x 5 x 10 ft.) on the roof of the Morrill Science Center at the University of Massachusetts. I was particularly anxious to see whether moths which escaped from the attacks of these captive birds might exhibit beak-damage patterns similar to those found on wild-caught specimens. It seemed likely that beak-damaged specimens obtained under circumstances where the behaviors of both birds and moths could be observed would permit some interpretations regarding similar specimens obtained from nature, and so might lead to hypotheses regarding the anti-predator significance of the hindwings.

The Blue Jays used in this study were one to two years old and had been hand-reared by Dr. Alan C. Kamil and his graduate students in the psychology department. These birds had been tested in a number of learning experiments (e.g., Hunter & Kamil, 1971), but their prior experience with insects was limited to mealworms (*Tenebrio* larvae), which had been used as rewards in these experiments, and to various flies which frequented their housing quarters.

Figure 8.1 A "classical" example of beak-damage in the Lepidoptera. Such tears from the vicinity of small eye-spots at the margin of butterfly wings are frequently encountered, and provide evidence for a deflection function of these eye-spots.

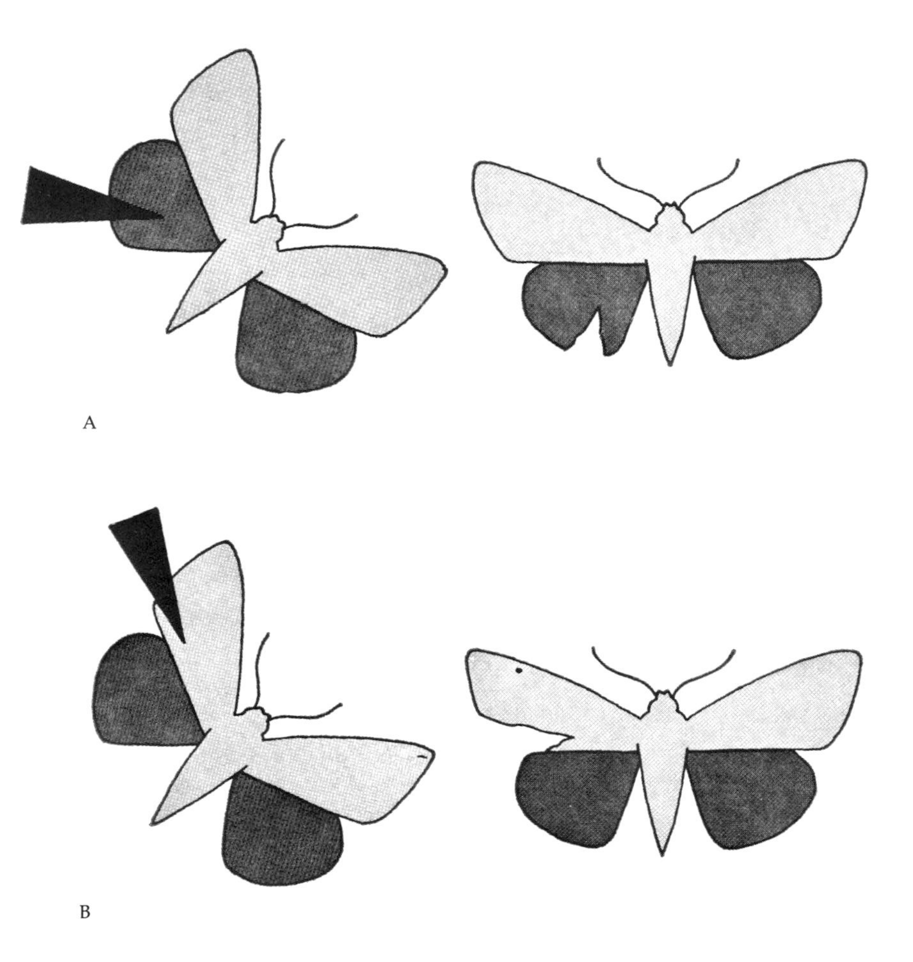

Figure 8.2 Type I damage patterns. Diagrammatic representations of bird attacks on flying moths (*left*) and the resulting specimens (*right*). Damage is usually confined to one hindwing (A), rarely to one forewing (B).

The *Catocala* released to these birds were fresh, undamaged specimens taken from my Robinson trap in Leverett, Massachusetts. The moths were released individually into one or the other of the two aviaries (each housing seven birds). Observation of subsequent moth-bird interactions included careful attention to such matters as: (1) whether a moth was attacked while flying or resting; (2) if attacked while resting, the direction from which attack occurred (rear, side, etc.); and (3) if escaping from attack, the duration of retention in the bird's beak. Of the 50 *Catocala* specimens released to the birds, 27 were successfully attacked and eaten, while 23 were attacked, but escaped and were recovered for later analysis of their beak-damage patterns. The birds became progressively more efficient at taking the moths, losing 10 of the first 12 released, but successfully capturing each of the last 15 specimens. Moths that were eaten by the birds were initially ingested wings and all, but after some experience the birds invariably attempted to remove the wings by holding the moths in their feet and pecking with their beaks.

The observations in these aviaries, coupled with analyses of the specimens which were retrieved following their attack by the Blue Jays, permitted a classification of beak-damage patterns into three major types: I (attacked while flying, tear from one wing), II (attacked while resting, tear from forewing and hindwing on the same side of the moth), III (attacked while resting, beak-imprint on at least one forewing). Each of these major categories has subdivisions, and the entire classification follows:

Type I. *Characteristic damage*: unilateral; tear from one wing only
a. hindwing tear (figs. 8.2A; 8.5A, B, C; 8.8A)
b. forewing tear (figs. 8.2B; 8.5D)

Type II. *Characteristic damage*: corresponding tears from ipsilateral forewing and hindwing
a. forewing and hindwing tears overlapping when wings *fully* closed; unilateral (figs. 8.3A; 8.6A, B, C)
b. forewing and hindwing tears overlapping when wings *partially* closed; unilateral (figs. 8.3B; 8.6D; 8.8B)
c. forewing and hindwing tears overlapping on both sides when wings *fully* closed; bilateral (figs. 8.3C; 8.6E, F)

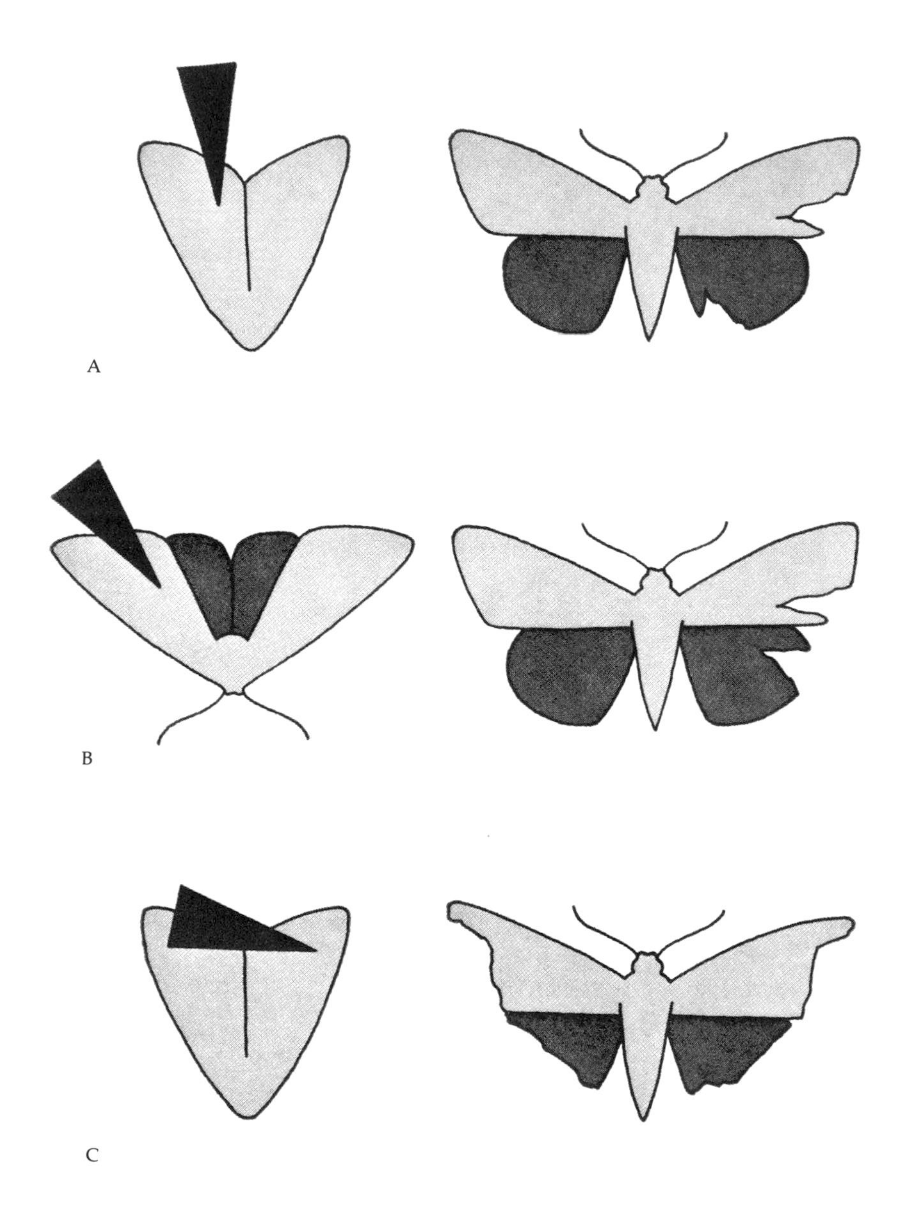

Figure 8.3 Type II damage patterns. Diagrammatic representations of bird attacks on resting moths (*left*) and the resulting specimens (*right*). Damage may be inflicted when the wings are fully closed (A) or partially open (B), and while usually unilateral, may be bilateral (C) on occasion.

Type III. *Characteristic damage:* crisp beak-imprint on one forewing; imprint or tear from ipsilateral hindwing

a. apex of beak-imprint directed toward (but not across) inner margin of forewing; unilateral (figs. 8.4A; 8.7A, B; 8.8A, B)

b. apex of beak-imprint directed toward (and usually across) costal margin of forewing; unilateral or bilateral (figs. 8.4B; 8.7C)

c. apex of beak-imprint directed toward (and across) inner margin of forewing; bilateral (figs. 8.4C; 8.7D)

This classification of beak-damage patterns, based on specimens obtained from the aviary sample, also proved adequate for classifying the specimens which had been obtained in the field. This field sample consisted of 65 specimens which were retained from a total of 1623 *Catocala* carefully examined at my collecting site in Leverett. (Ten additional specimens which were originally suspected as beak-damaged proved upon subsequent study to have bat-inflicted damage patterns — see fig. 8.9B).

The complete data on the damage patterns obtained in both the aviary and field samples are given in table 8.1. Note that a number of specimens (6 in the aviary sample, 8 in the field sample) exhibited two damage patterns (e.g., fig. 8.8A,B). Analysis of these data provides evidence that *Catocala* hindwings serve both deflective and startle functions.

Nearly half of the beak-damaged specimens in both samples were attacked while in flight, and all but one of these exhibited only hindwing damage. It appears that bird attacks are frequently directed towards these structures when *Catocala* moths are flying. I have witnessed a number of birds attacking *Catocala* in flight, and despite hearing the snap of their beaks, I have seen the moths successfully reach a tree trunk; many birds (e.g., Crested Flycatcher [*Myiarchus crinitus*], Catbird [*Dumetella carolinensis*], Scarlet Tanager [*Piranga olivacea*]) then terminate their attacks. Such chases may result in damage to the moths' hindwings, but such damage is probably not highly detrimental (some specimens reach my Robinson trap with one hindwing completely missing — fig. 8.9A). *Catocala* hindwings apparently function on some occasions as deflective devices, directing bird attacks away from more vulnerable body parts.

The remaining half of the beak-damaged specimens were attacked while resting. Escape from these attacks seemed to be of

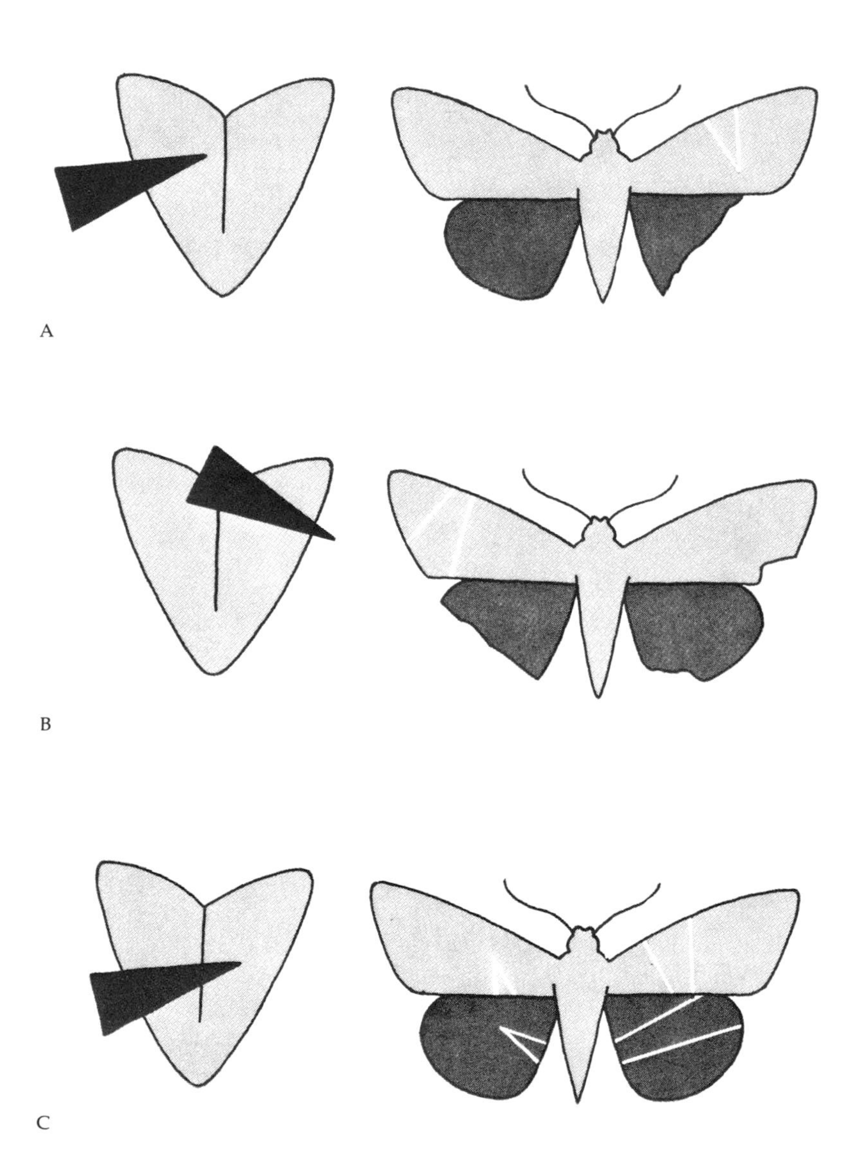

Figure 8.4 Type III damage patterns. Diagrammatic representations of bird attacks on resting moths (*left*) and the resulting specimens (*right*). The beak-mark is usually confined to one forewing, including an imprint of the apex (A) or not (B), but may extend across both forewings (C).

two sorts: (1) a pulling-free of the moth *while* being tightly gripped by a bird, resulting in a tearing of the wings around the region of beak contact (Type II damage), and (2) a release of the moth *after* being tightly gripped by a bird, resulting in a clear beak-imprint along some of the lines of beak contact (Type III damage).

Type II damage apparently occurred when the size, speed, and strength of the moth enabled it to break free from the grip of a bird. This may be a common occurrence when a small bird attempts to take a large *Catocala*. I once observed a persistent White-breated Nuthatch (*Sitta carolinensis*) repeatedly attempt to capture large *Catocala* on tree trunks in the vicinity of my Robinson trap, and whenever the moths were actually grasped by the bird, they were able to break free quickly.

It is postulated that Type III damage was the result of startle responses of the predators to the sudden appearance of the contralateral hindwing (i.e., opposite the side being gripped). The startle response is viewed as effecting a momentary relaxation of a bird's grip, enabling the moth to escape but leaving a crisp beak-imprint on some part of its wings (plate VIII). (It seems likely that some of the specimens exhibiting Type II damage could also have startled their predators, but that tears in the wings occurred or developed because the bird's grip included only the more fragile portions of the forewing.)

The only damage pattern not obtained in the field sample was Type IIIc, the only pattern involving an attack during which the contralateral hindwing cannot be displayed. This attack (fig. 8.4c, left) was the one that Blue Jays in the aviaries used almost exclusively after they had had some experience with *Catocala*. A moth grasped in this fashion rarely escaped, but on one occasion a bird apparently loosened its grip when attempting to transfer a moth from its beak to its feet, and the escaped specimen was recovered (fig. 8.7D). Presumably such "carelessness" would be very rare in nature.

A contradiction may appear between the proposed functions of deflection and startle for the same structures with respect to the same predators. Resolution of the difficulty could lie in an understanding of the nature of startle in this situation. I suggest that the degree of startle exhibited by a bird in any encounter with a *Catocala* hindwing pattern is not so closely related to that pattern per se, as to the degree of expectation of that pattern which the

Figure 8.5 Specimens exhibiting Type I damage patterns: (A) *C. antinympha,* IA (Robinson trap); (B) *C. retecta,* IA (experimental moth); (C) *C. unijuga,* IA (bait); (D) *C. residua,* IB (black light). 0.8X.

Figure 8.6 Specimens exhibiting Type II damage patterns: (A) *C. ilia*, IIA (bait); (B) *C. epione*, IIA (black light); (C) *C. dejecta*, IIA (Robinson trap); (D) *C. flebilis*, IIB (Robinson trap); (E) *C. retecta*, IIC (Robinson trap); (F) *C. concumbens*, IIC (Robinson trap). 0.8X.

Figure 8.7 Specimens exhibiting Type III damage patterns: (A) *C. ultronia,* IIIA (experimental moth); (B) *C. retecta,* IIIA (experimental moth); (C) *C. ultronia,* IIIB (Robinson trap); (D) *C. concumbens,* IIIC (experimental moth). 0.8X.

bird brings to the encounter on the basis of its past experience. Thus, *anomaly,* defined in terms of departure from expectation, is regarded as the critical factor in determining whether a bird will react by releasing a *Catocala* moth which suddenly displays a particular hindwing pattern. This instantaneous reaction might not interfere in any substantial way with the ability of a bird to attack a flying moth, if time to adjust to the appearance of the hindwings was available. Indeed, some specimens exhibiting both Type I and Type III damage (e.g., fig. 8.8A) may demonstrate the successive utilization of the startle and deflective functions of the hindwings with respect to the same individual predator.

An "anomalous stimulus" (defined in terms of its departure from expectation and the momentary startle it elicits) would be very similar, though perhaps not identical, to what has been referred to as a "protean stimulus" (defined in terms of its unpredictability and the confusion it elicits), and to a "novel stimulus" (defined in terms of its unfamiliarity and the avoidance it elicits). It seems likely that these various stimuli, whatever their definitional differences, may produce similar physiological effects (e.g., arousal or conflict; Sokolov, 1960; Coppinger, 1970; Driver & Humphries, 1970) which result in reduced predator efficiency.

The primary evidence that *Catocala* hindwings function as anomalous stimuli is provided by specimens which exhibit crisp beak-imprints on their wings (Type III damage). Such beak-marks on lepidopteran wings have long been regarded as evidence that the specimens involved were captured and subsequently released by birds. Since beak-marks of this sort are most often found on warningly colored species, it has been assumed that release of the specimens followed predator recognition of some noxious quality (usually odor or taste) of the captured prey. However, *Catocala* moths, as far as is known, are entirely palatable, and their release by birds must be related to some sort of startle response. Again, it is postulated that interspecific hindwing variation provides the key to understanding this situation. Hindwing variation introduces the potential of anomalous stimuli into the predator-prey relationship. Thus, a predator is seen as building up expectations regarding future hindwing patterns on the basis of its past experiences with these patterns, reacting more and more efficiently if these expectations are met, but inefficiently if they are countered. Inefficiency presumably

Figure 8.8 Specimens exhibiting two damage patterns: (A) *C. ultronia,* IIIA on the left side, IA on the right side (black light); (B) *C. ultronia,* IIB on the left side, IIIA on the right side (Robinson trap). 1.3X.

results from some involuntary response (e.g., gaping) elicited by an unexpected stimulus which interferes with the completion of normal attack, allowing a moth to escape.

Some indirect evidence for this view is provided by an analysis of the distribution of beak-damage patterns on moths having different hindwing types. The most obvious discontinuity among *Catocala* hindwings occurs between the chromatic (colored) and achromatic (lacking color) patterns. As individuals possessing chromatic hindwings are more common than those possessing achromatic hindwings (the latter comprising less than 25% of the *Catocala* taken in Leverett each year), it might be predicted that achromatic hindwings would be less often encountered, and therefore more often anomalous, than chromatic hindwings. Analysis of the data in table 8.2 reveals that beak-damage patterns II and III are more commonly found on individuals possessing achromatic hindwings than would be expected on the basis of chance. This finding suggests that achromatic hindwings are particularly effective as startle devices, and anomaly may provide an explanation for this effectiveness.

Predators exhibiting a tendency to react inefficiently to anomalous stimuli would provide strong selection pressures favoring diversity in their prey (apostatic selection). Diversity would result in an increased number of predator expectations and a corresponding increased number of potential anomalous stimuli. Anomaly, within the limits of the advantage it provides, would then favor the origin and maintenance of considerable diversity in sympatric assemblages of closely related species.

Recently, Dr. Denis Owen and I attempted an analysis of hindwing diversity in four large *Catocala* samples taken at four localities in eastern North America (Sargent & Owen, 1975). In each case, all of the *Catocala* taken at a mercury vapor lamp over at least one entire season were recorded. For purposes of our analysis, the hindwing patterns were arbitrarily divided into five groups: (1) black and white, banded; (2) black, unbanded (on upper surface); (3) yellow to yellow-orange and black, banded; (4) orange-red to red and black, banded; and (5) pink and black, banded. The division of species into these hindwing groups and their frequencies at each of the four localities are given in table 8.3. It is immediately apparent that the species composition was

Table 8.1 Distribution of damage patterns among *Catocala* in an aviary and field sample.

Damage patterns		Number of individual examples	
		Aviary sample	Field sample
Type I	a	12	33
	b	1	1
Type II	a	5	16
	b	2	3
	c	1	2
Type III	a	5	13
	b	2	5
	c	1	0
Totals		29	73

Table 8.2 Distribution of damage patterns among field-caught *Catocala* of two hindwing types.

Hindwing types	Number of individual examples			
	Damage patterns			
	I	II	III	Total sample
Chromatic	28	14	10	1228
Achromatic	6	7	8	395
Percent Achromatic	17.6	33.3	44.4	24.3

Table 8.3 Number of *Catocala* moths in five hindwing groups obtained at mercury vapor lamps.

Hindwing Group	Species	Leverett, Mass. (1970–73)	W. Hatfield, Mass. (1970–73)	Washington, Conn. (1970–73)	George Reserve, Mich. (1961)
1.	*relicta*	5	7	7	11
2.	*epione*	163	144	59	167
	judith	20	20	166	1
	flebilis	22	5	18	—
	obscura	3	15	82	—
	residua	66	94	298	3
	retecta	278	258	96	43
	dejecta	19	3	4	52
	vidua	—	—	—	2
	andromedae	87	42	58	—
3.	*piatrix*	—	2	—	1
	antinympha	339	197	88	—
	badia	—	—	60	—
	habilis	212	304	259	50
	serena	—	—	213	13
	palaeogama	182	250	886	58
	subnata	1	5	10	—
	neogama	28	99	98	16
	cerogama	1	16	—	8
	gracilis	141	38	33* (gracilis + sordida)	19
	sordida	160	45		18
	crataegi	55	44* (crataegi + mira + blandula)	38	—
	mira	18		129	2
	blandula	46		23	—
	grynea	184	63	181	5
	praeclara	67	10	20	9
	similis	52	5	—	38
	micronympha	65	45	32	—
	connubialis	8	21		—
	amica	279	714	388	500
4.	*innubens*	—	1	1	—
	ilia	50	324	25	185
	parta	1	22	11	8
	briseis	—	1	1	2
	meskei	—	—	—	1
	unijuga	19	29	9	5
	coccinata	53	22	19	20
	ultronia	348	184	382	3
5.	*cara*	4	18	34	6
	concumbens	40	76	322	84
	amatrix	—	42	5	1
Total	Individuals	3016	3165	4055	1331
	Species	33	37	35	30

*Species not differentiated

Figure 8.9 Specimens exhibiting extreme damage: (A) *C. epione*, left hindwing missing, presumably result of bird attack similar to that which produced specimen in figure 8.6C (Robinson trap); (B) *C. ilia*, bat-inflicted damage, resulting in tattered wing margins on reasonably fresh specimen (Robinson trap); (C) *C. judith*, probably weathering damage only, as specimen very late (22 September) and worn (Robinson trap); (D) *C. ilia*, bird damage (Type I) and weathering, specimen very late (22 September) (bait). 0.8X.

very different at the four localities, presumably because of differences in larval foodplant availability (fig. 8.10B), but that hindwing diversity was remarkably similar at each of the locations (fig. 8.10A).

The apparent stability in hindwing diversity suggests that particular hindwing patterns may be advantageous only if they do not exceed a certain percentage of the total of hindwings present at any locality. Perhaps the frequency at which a particular pattern occurs is determined in part by the requirements of anomaly. For example, achromatic hindwings (groups 1 and 2) comprised approximately 20% of the total at each of the four localities. If this frequency were exceeded, the probability of achromatic hindwings being anomalous would likely be substantially reduced. In that event, predation might become more effective on moths having achromatic hindwings, and this would tend to bring the frequency back to its original level.

The *Catocala* samples just discussed represented summations of captures over one or more entire seasons. It is also of interest to look at the frequencies of occurrence of the various hindwing patterns at different times within one season, and such data are presented in fig. 8.11 for 1971 in Leverett, Massachusetts. These data reveal that the various hindwing types have rather different patterns of seasonal occurrence. Black hindwings, for example, become progressively more common over the season, while red and black banded hindwings become less common as the season progresses. Such changes in the frequencies of occurrence of different hindwing patterns should serve to further confound predators as they build up expectations regarding the startle patterns of the *Catocala*.

Several lines of evidence suggest, therefore, that hindwing diversity in the *Catocala* is a response to pressures exerted by predators, and that it functions to reduce the efficacy of predation by introducing anomalous stimuli which are particularly effective as startle devices. Anomaly can only be achieved in a system which possesses sufficient order to encourage expectations in the predators, and this requirement may explain such order as we have seen in the frequencies and seasonal occurrence of the various hindwing patterns.

If any thread runs through this chapter, and indeed through this book, it is the recurring emphasis on diversity in the *Catocala*. Diversity is not only a measure of success in these moths, but also a key to understanding that success.

Catocala diversity, though extensive, is confined within the limits of an overall unity. Like a composer constrained by a theme, the *Catocala* are limited to variations which conform to certain constant principles of design. The forewings must be cryptic, but can vary within species; the hindwings must be boldly patterned, and must not vary within species. Given such restrictions, the *Catocala*, like a skillful composer, have developed the most marvelously complex variations.

All of this diversity must be intended for the eyes of some

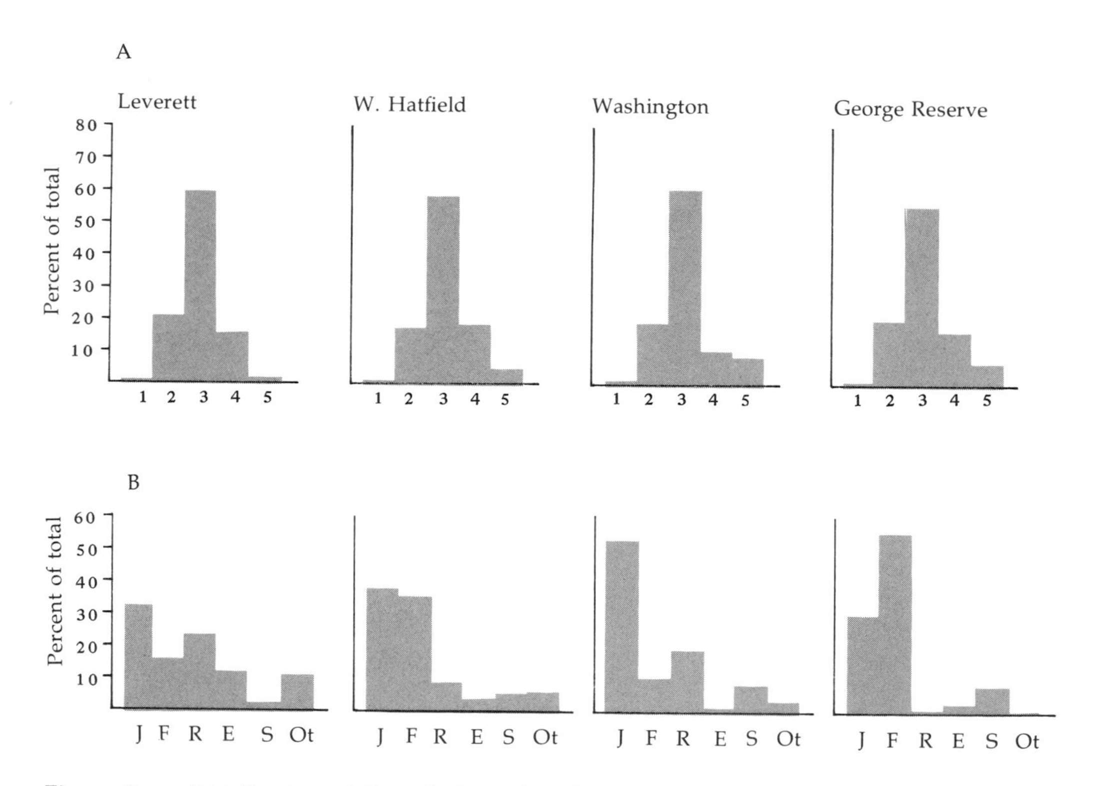

Figure 8.10 Distribution of *Catocala* from four locations (table 8.3) according to hindwing patterns (A) and foodplant families (B). The five hindwing groups are described in the text. Foodplant families: J (Juglandaceae), F (Fagaceae), R (Rosaceae), E (Ericaceae), S (Salicaceae), Ot (others).

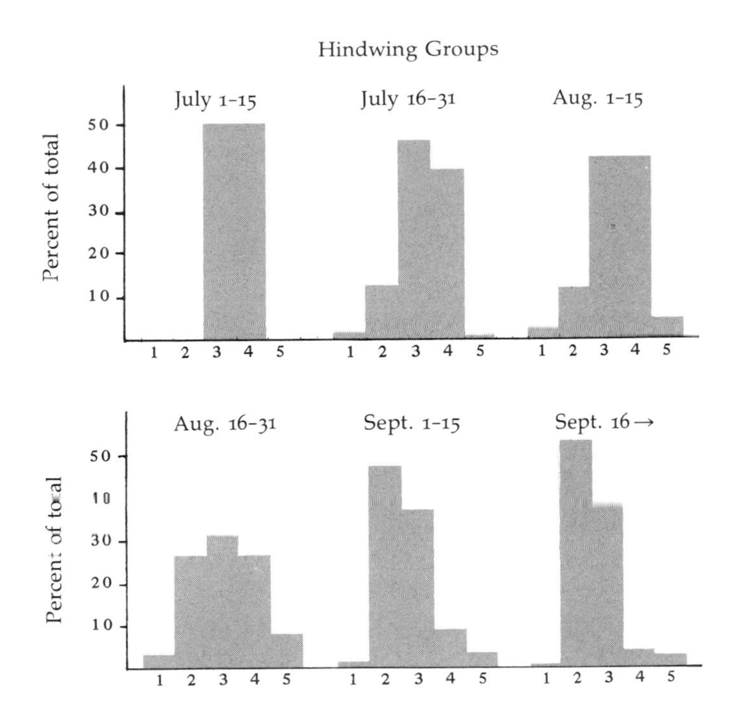

Figure 8.11 Distribution of hindwing patterns in *Catocala* collected over successive two-week periods during the season of 1971 in Leverett, Massachusetts. The five hindwing groups are described in the text.

beholders, and the evidence suggests that these beholders are birds—birds, beautiful but certainly not brilliant, and consummate conservatives, especially in matters of diet. Diversity would seem the perfect bane of birds! For as far as a bird is concerned, if an item is not tried and true, it (1) doesn't exist, (2) is frightening, or (3) should be ignored. Though birds may learn when relationships are simple and systematic, diversity makes for complex and unpredictable relationships which only confound the avian brain.

So in the world of the *Catocala*, diversity is at a premium, and this fact may explain the extraordinary number of species which characterizes this genus. In any group of *Catocala* species, a new species should be welcome as long as the advantage it brings in terms of additional diversity is greater than any disadvantage associated with its competition for resources. In this way, assuming plentiful resources, one may envision the development of large, sympatric assemblages of these moths.

Appendix 1

Collecting Data — Leverett, Massachusetts

The following is a record of all *Catocala* individuals taken from 1968 to 1973. The data for each species are broken down by year and collecting procedure. Abbreviations are:

- RT (Robinson mercury vapor light-trap, 125-watt Mazda bulb)
- BL (fluorescent black light, 15-watt General Electric bulb, F15T8-BL)
- SP (incandescent spotlights, 150-watt Westinghouse and General Electric outdoor bulbs)
- BT (bait, usually brown sugar and beer)
- RE (resting, in a natural site)

The species are arranged essentially as in the McDunnough checklist (1938).

In the records for each species, blanks indicate that a procedure was not utilized during the year in question, while dashes indicate that the species was not taken though the procedure was utilized. If a species was entirely missed in any year, that year is omitted from the records.

Species	Collecting procedures					
Years	RT	BL	SP	BT	RE	Totals
piatrix						
1971	—	—	—	1	—	1
epione						
1968			4	3		7
1969		1	3	5		9
1970	70	16	12	2	3	103
1971	70	20	4	19	5	118
1972	7	12	7	—	4	30
1973	16	3	1		1	21
Totals	163	52	31	29	13	288
antinympha						
1968			6	—		6
1969		4	7	4		15
1970	98	11	8	—	3	120
1971	140	13	5	5	4	167
1972	47	3	4	—	—	54
1973	54	4	4		—	62
Totals	339	35	34	9	7	424
habilis						
1969		1	—	—		1
1970	98	15	5	—	3	121
1971	75	1	—	—	2	78
1972	32	3	3	—	2	40
1973	7	2	—		—	9
Totals	212	22	8	—	7	249

Species	Collecting procedures					
Years	RT	BL	SP	BT	RE	Totals
judith						
1970	7	—	—	—	—	7
1971	6	1	—	—	1	8
1972	2	—	—	—	—	2
1973	5	—	—		—	5
Totals	20	1	—	—	1	22
flebilis						
1970	10	—	—	—	—	10
1971	3	—	—	—	—	3
1972	8	—	—	—	—	8
1973	1	—	—		—	1
Totals	22	—	—	—	—	22
obscura						
1969		1	—	—		1
1970	2	1	—	—	1	4
1971	1	—	—	—	—	1
1972	—	1	—	—	—	1
Totals	3	3	—	—	1	7
residua						
1969		2	—	—		2
1970	21	—	1	—	1	23
1971	34	2	2	3	1	42
1972	11	4	2	—	2	19
1973	—	—	—		1	1
Totals	66	8	5	3	5	87
retecta						
1968			2	5		7
1969		6	1	11		18
1970	99	16	4	3	4	126
1971	143	18	14	25	2	202
1972	23	17	4	—	1	45
1973	13	3	1		—	17
Totals	278	60	26	44	7	415
dejecta						
1969		—	1	—		1
1970	6	—	—	—	—	6
1971	6	—	—	1	—	7
1972	2	1	—	—	1	4
1973	5	—	—		—	5
Totals	19	1	1	1	1	23

Species	Collecting procedures					
Years	RT	BL	SP	BT	RE	Totals
palaeogama						
1968			1	—		1
1969		1	1	—		2
1970	108	5	2	—	4	119
1971	47	5	1	—	—	53
1972	13	—	—	—	—	13
1973	14	—	—			14
Totals	182	11	5	—	4	202
subnata						
1973	1	—	—		—	1
neogama						
1970	20	1	—	—	3	24
1971	7	1	2	1	1	12
1972	1	1	—	—	—	2
Totals	28	3	2	1	4	38
ilia						
1968			6	209		215
1969		5	3	101		109
1970	20	12	1	47	3	83
1971	24	14	7	330	1	376
1972	3	3	—	17	1	24
1973	3	—	—		—	3
Totals	50	34	17	704	5	810
cerogama						
1968			—	1		1
1971	—	—	—	2	—	2
1972	1	—	—	—	—	1
Totals	1	—	—	3	—	4
relicta						
1968			—	1		1
1969		—	—	6		6
1970	3	2	—	1		6
1971	2	1	2	26	3	34
1972	—	—	—	—	1	1
1973	—	1	—		—	1
Totals	5	4	2	34	4	49
unijuga						
1968			—	6		6
1969		1	—	4		5
1970	11	—	—	—	5	16

Species	Collecting procedures					
Years	RT	BL	SP	BT	RE	Totals
1971	7	—	3	18	—	28
1972	1	2	—	—	1	4
1973	—	1	—		—	1
Totals	19	4	3	28	6	60
parta						
1970	1	1	—	—	1	3
cara						
1968			—	3		3
1969		1	—	8		9
1970	2	—	1	26	8	37
1971	2	2	2	13	3	22
1972	—	1	—	—	1	2
1973	—	—	—	1	1	2
Totals	4	4	3	51	13	75
concumbens						
1968			1	4		5
1969		1	—	9		10
1970	18	1	1	—	3	23
1971	15	5	4	18	—	42
1972	3	2	—	—	—	5
1973	4	1	—		—	5
Totals	40	10	6	31	3	90
amatrix						
1969		—	—	1		1
1970	—	—	1	—	—	1
1971	—	—	—	1	1	2
1972	—	—	—	—	2	2
Totals	—	—	1	2	3	6
*sordida**						
1970	36	4	6	—	—	46
1971	79	9	5	3	—	96
1972	28	3	—	—	—	31
1973	17	—	1		1	19
Totals	160	16	12	3	1	192
*gracilis**						
1970	63	19	7	—	—	89
1971	39	12	7	7	—	65
1972	12	6	2	—	—	20
1973	27	1	4		—	32
Totals	141	38	20	7	—	206

**sordida* and *gracilis* were not distinguished prior to 1970

Species	Collecting procedures					
Years	RT	BL	SP	BT	RE	Totals
andromedae						
1968			—	3		3
1969		—	2	2		4
1970	30	3	7	3	—	43
1971	33	6	3	11	2	55
1972	13	2	3	—	—	18
1973	11	—	—		—	11
Totals	87	11	15	19	2	134
coccinata						
1968			1	—		1
1969		—	1	—		1
1970	8	4	1	—	—	13
1971	27	4	1	—	2	34
1972	9	2	—	—	—	11
1973	9	1	—		—	10
Totals	53	11	4	—	2	70
ultronia						
1968			4	44		48
1969		2	6	31		39
1970	162	11	6	2	4	185
1971	124	7	8	99	1	239
1972	21	3	1	—	—	25
1973	41	—	2		1	44
Totals	348	23	27	176	6	580
*crataegi***						
1970	9	1	2	—	—	12
1971	12	2	2	6	—	22
1972	20	3	—	—	—	23
1973	14	2	—		—	16
Totals	55	8	4	6	—	73
*mira***						
1970	7	3	—	—	—	10
1971	7	—	—	1	—	8
1972	4	1	—	—	—	5
Totals	18	4	—	1	—	23
*blandula***						
1970	18	1	2	—	1	22
1971	19	4	2	26	—	51
1972	8	—	1	—	—	9
1973	1	1	1		—	3
Totals	46	6	6	26	1	85

***crataegi, mira* and *blandula* were not distinguished prior to 1970

Species	Collecting procedures					
Years	RT	BL	SP	BT	RE	Totals
grynea						
1968			—	3		3
1969		1	—	1		2
1970	101	9	3	—	—	113
1971	46	4	—	7	—	57
1972	19	1	—	—	—	20
1973	18	1	1		—	20
Totals	184	16	4	11	—	215
praeclara						
1968			—	2		2
1969		1	—	3		4
1970	22	—	1	—	1	24
1971	22	4	3	10	2	41
1972	10	—	—	—	—	10
1973	13	—	—		—	13
Totals	67	5	4	15	3	94
similis						
1970	9	3	2	—	—	14
1971	11	—	1	3	—	15
1972	16	1	—	—	—	17
1973	16	—	—		—	16
Totals	52	4	3	3	—	62
micronympha						
1969		1	2	—		3
1970	17	9	5	1	—	32
1971	27	8	9	2	2	48
1972	11	10	5	—	—	26
1973	10	5	3		—	18
Totals	65	33	24	3	2	127
connubialis						
1970	—	—	1	—	—	1
1971	3	—	—	—	—	3
1972	3	—	1	—	—	4
1973	2	1	—	1	—	4
Totals	8	1	2	1	—	12
amica						
1969		1	—	—		1
1970	85	21	12	—	1	119
1971	83	22	3	9	—	117
1972	34	28	7	—	—	69
1973	77	5	1		—	83
Totals	279	77	23	9	1	389

Appendix 2

Sex Ratios — Leverett, Massachusetts

The numbers of males and females of *Catocala* taken from 1968 to 1973 are given here. The data for each species are broken down by collecting procedure, though summed across years. Abbreviations for collecting procedures are the same as in Appendix 1.

	Males/females (Percent male)				
Species	RT	BL	SP	BT	RE
piatrix	—	—	—	1/— (100)	—
epione	103/13 (89)	27/22 (55)	4/19 (17)	2/14 (13)	5/4 (56)
antinympha	156/19 (89)	13/16 (45)	3/15 (17)	5/ (100)	4/1 (80)
habilis	181/13 (93)	13/8 (62)	5/3 (63)	—	4/1 (80)
judith	19/1 (95)	—/1 (0)	—	—	1/— (100)
flebilis	18/3 (86)	—	—	—	—
obscura	3/— (100)	2/— (100)	—	—	1/— (100)
residua	56/2 (97)	6/— (100)	4/— (100)	3/— (100)	3/— (100)
retecta	220/17 (93)	23/31 (43)	11/12 (48)	15/11 (58)	3/2 (60)
dejecta	19/— (100)	1/— (100)	—	—/1 (0)	1/— (100)
palaeogama	162/1 (99)	8/2 (80)	3/— (100)	—	1/— (100)
subnata	1/— (100)	—	—	—	—
neogama	24/1 (96)	2/1 (67)	1/1 (50)	1/— (100)	2/1 (67)
ilia	35/11 (76)	16/11 (59)	6/2 (75)	109/239 (31)	2/— (100)
cerogama	1/— (100)	—	—	—/2 (0)	—
relicta	4/1 (80)	3/1 (75)	—/1 (0)	9/14 (39)	2/1 (67)
unijuga	16/2 (89)	3/— (100)	1/2 (33)	9/8 (53)	3/— (100)
parta	—	1/— (100)	—	—	1/— (100)

Species	Males/females (Percent male)				
	RT	BL	SP	BT	RE
cara	3/1 (75)	2/1 (67)	2/1 (67)	18/12 (60)	10/1 (91)
concumbens	37/— (100)	7/1 (88)	2/3 (40)	11/3 (79)	1/1 (50)
amatrix	—	—	—/1 (0)	1/— (100)	—/3 (0)
sordida	94/9 (91)	7/9 (44)	4/8 (33)	1/2 (33)	1/— (100)
gracilis	96/6 (94)	12/26 (32)	3/17 (15)	4/3 (57)	—
andromedae	61/4 (94)	1/10 (9)	—/12 (0)	2/9 (18)	—/1 (0)
coccinata	47/2 (96)	4/7 (36)	2/— (100)	—	1/— (100)
ultronia	217/11 (95)	10/11 (48)	7/9 (44)	73/21 (78)	4/— (100)
crataegi	35/1 (97)	1/5 (17)	—/4 (0)	3/2 (60)	—
mira	16/— (100)	1/3 (25)	—	1/— (100)	—
blandula	35/1 (97)	5/1 (83)	1/5 (17)	16/7 (70)	1/— (100)
grynea	105/4 (96)	8/7 (53)	1/3 (25)	4/1 (80)	—
praeclara	48/2 (96)	3/1 (75)	1/3 (25)	6/3 (67)	1/— (100)
similis	26/1 (96)	1/3 (25)	1/2 (33)	1/2 (33)	—
micronympha	46/2 (96)	19/12 (61)	7/14 (33)	—/2 (0)	2/— (100)
connubialis	7/1 (88)	—/1 (0)	1/— (100)	1/— (100)	—
amica	144/16 (90)	49/25 (66)	21/1 (95)	2/6 (25)	1/— (100)
Totals	2035/145 (93)	248/216 (53)	91/138 (40)	298/362 (45)	55/16 (77)

Appendix 3

Collecting Data — West Hatfield, Massachusetts

These records are the numbers of *Catocala* taken from 1969 to 1973, at bait (brown sugar and red cooking wine) and a Robinson mercury vapor light-trap (abbreviated BT and RT, respectively). In 1971, collecting did not commence until mid-August; this fact undoubtedly explains the absence of early flying *Catocala* in that year. The original data were obtained by Charles G. Kellogg, who kindly made them available for inclusion here.

	Numbers of individuals						
	1969	1970		1971	1972	1973	
Species	BT	RT	BT	RT	RT	RT	Totals
innubens	—	—	—	1	—	—	1
piatrix	1	—	1	1	1	—	4
epione	6	55	7	14	23	52	157
antinympha	6	27	2	25	30	115	205
habilis	2	87	1	133	62	22	307
judith	—	1	—	10	2	7	20
flebilis	—	1	—	4	—	—	5
obscura	—	6	—	7	1	1	15
residua	9	18	2	56	7	13	105
retecta	34	97	19	131	20	10	311
dejecta	—	—	—	—	1	2	3
palaeogama	6	81	3	134	21	14	259
subnata	—	—	—	2	1	2	5
neogama	4	19	—	65	8	7	103
ilia	223	149	78	101	42	32	625
cerogama	7	4	1	5	6	1	24
relicta	4	2	1	3	1	1	12
unijuga	5	2	—	17	7	3	34
parta	1	5	—	9	7	1	23
briseis	—	—	—	1	—	—	1
cara	50	2	36	6	6	4	104
concumbens	19	12	1	33	20	11	96
amatrix	34	6	28	24	7	5	104
sordida	—	30	—	4	5	6	45

	Numbers of individuals						
	1969	1970		1971	1972	1973	
Species	BT	RT	BT	RT	RT	RT	Totals
gracilis	—	15	—	6	4	13	38
andromedae	—	18	1	16	—	8	43
coccinata	—	5	—	3	4	10	22
ultronia	37	52	8	42	25	65	229
crataegi / *mira* / *blandula* *	2	8	2	—	9	27	48
grynea	7	35	1	1	5	22	71
praeclara	1	7	—	—	3	—	11
similis	—	3	—	—	—	2	5
micronympha	—	15	—	—	11	19	45
connubialis	—	1	—	—	4	16	21
amica	1	200	5	147	101	266	720

*These three species were not always distinguished.

Appendix 4

Collecting Data — Washington, Connecticut

These records give the numbers of *Catocala* taken each year from 1961 to 1965 and 1967 to 1973, and are entirely from two light sources: a Robinson mercury vapor light-trap (most records), and a 15-watt fluorescent blacklight. The original data were obtained by the late Sidney A. Hessel, who kindly provided all of his records for these years.

	Numbers of individuals												
Species	1961	1962	1963	1964	1965	1967	1968	1969	1970	1971	1972	1973	Totals
innubens	1	—	—	—	2	1	—	—	—	1	1	—	6
piatrix	1	—	—	—	—	—	—	—	—	—	—	—	1
epione	69	37	14	6	19	24	21	15	16	26	5	12	264
antinympha	52	26	13	13	16	182	24	27	18	18	27	25	441
badia	9	7	3	2	8	21	10	7	11	27	9	13	127
habilis	121	79	63	119	78	159	52	34	97	95	44	23	964
serena	3	—	—	8	12	52	37	28	78	124	8	3	353
judith	1	—	—	—	1	9	10	24	45	95	12	14	211
flebilis	10	2	1	—	1	4	3	1	14	3	1	—	40
obscura	27	17	3	17	9	26	18	25	27	28	19	8	224
residua	216	345	45	53	54	135	32	39	56	207	28	7	1217
retecta	83	55	19	29	34	55	13	19	31	39	17	9	403
dejecta	10	7	1	2	3	15	1	1	2	4	—	—	46
vidua	—	—	—	1	—	—	—	—	—	—	—	—	1
palaeogama	194	487	12	40	33	81	14	48	144	690	29	23	1795
subnata	9	19	2	2	1	1	—	—	—	7	1	2	44
neogama	97	36	23	38	39	68	19	28	49	10	4	5	446
ilia	9	4	1	5	9	4	5	5	4	7	1	13	67
cerogama	—	—	—	—	—	1	—	—	—	—	—	—	1
relicta	6	—	1	1	5	2	—	1	3	4	—	—	23
unijuga	7	5	3	6	4	8	—	—	—	6	2	1	42
parta	9	2	6	3	4	7	1	2	5	2	4	—	45
briseis	—	—	—	—	—	2	—	—	—	1	—	—	3
cara	15	12	3	6	2	5	1	4	4	10	16	4	82
concumbens	14	12	21	59	68	75	45	60	45	130	99	48	676
amatix	—	1	1	—	—	—	—	—	—	1	3	1	7
sordida / *gracilis* *	16	7	12	15	10	7	6	8	10	15	6	2	114
andromedae	40	30	7	8	26	10	15	18	9	34	3	12	212
coccinata	7	4	1	2	1	2	1	2	1	11	7	—	39

Species	Numbers of individuals												Totals
	1961	1962	1963	1964	1965	1967	1968	1969	1970	1971	1972	1973	
ultronia	41	38	14	24	32	45	19	37	62	220	65	35	632
crataegi	1	2	1	1	—	2	1	1	5	19	11	3	47
mira	4	14	8	11	9	2	4	20	35	76	10	8	201
blandula	—	6	3	1	2	2	3	—	1	21	1	—	40
grynea	72	30	8	19	35	39	26	56	52	64	17	48	466
praeclara	1	—	1	1	4	6	1	4	3	11	1	5	38
similis	7	7	1	—	1	4	1	—	—	—	—	—	21
micronympha	38	43	5	11	4	6	5	15	9	20	1	2	159
amica	85	78	10	27	27	89	36	50	49	281	14	44	790
Totals	1275	1412	306	530	553	1151	424	579	885	2337	466	370	10,288
No. species	33	29	31	30	32	35	29	28	29	34	31	26	39

* These two species were not always distinguished.

Appendix 5

Data on Figured Specimens

Chapter II

Figure 2.1

A. (left). *C. ilia*, ♂. Batsto, Burlington Co., New Jersey; 6 July 1974; D. F. Schweitzer (DFS).
(right). *C. ilia*, ♀. Harwich, Barnstable Co., Massachusetts; July 1919; J. B. Paine (JBP).
B. (left). *C. lacrymosa* "evalina," ♂. Eureka, St. Louis Co., Missouri; 8 August 1937; F. R. Arnhold (PMNH).
(right). *C. lacrymosa* "paulina," ♂. Virginia? (USNM).
C. (left). *C. relicta* "clara," ♂. Leverettt, Franklin Co., Massachusetts; 11 August 1971; T. D. Sargent (TDS).
(right). *C. relicta* "clara," ♀. Leverett, Franklin Co., Massachusetts; ex ova, 1970; T. D. Sargent (TDS).
D. (left). *C. habilis*, ♂. Leverett, Franklin Co., Massachusetts; 20 August 1970; T. D. Sargent (TDS).
(right). *C. habilis*, ♀. West Hatfield, Hampshire Co., Massachusetts; 28 August 1970; C. G. Kellogg (CGK).
E. (left). *C. retecta*, ♂. West Hatfield, Hampshire Co., Massachusetts; 10 September 1970; C. G. Kellogg (CGK).
(right). *C. retecta*, ♀. West Hatfield, Hampshire Co., Massachusetts; 27 August 1970; C. G. Kellogg (CGK).

Figure 2.4

A. *C. andromedae*, ♀. Pelham, Hampshire Co., Massachusetts; 27 July 1964; T. D. Sargent (TDS).
B. *C. miranda*, ♀. Fontana Dam, Graham Co., North Carolina; 9 July 1972; D. F. Schweitzer (DFS).
C *C. epione*, ♀. Leverett, Franklin Co., Massachusetts; 26 July 1970; T. D. Sargent (TDS).
D. *C. judith*, ♂. Leverett, Franklin Co., Massachusetts; 26 July 1971; T. D. Sargent (TDS).
E. *C. robinsoni*, ♂. Missouri; 4 Sepptember 1932; (TDS).
F. *C. lacrymosa*, ♂. Missouri; 28 August 1932; (TDS).
G. *C. dejecta*, ♀. Leverett, Franklin Co., Massachusetts; 29 july 1969; T. D. Sargent (TDS).
H. *C. insolabilis*, ♂. Lusk Creek, Pope Co., Illinois; 23 June 1968; J. C. Downey (MCN).

Figure 2.5

A. C. *crataegi,* ♀. Leverett, Franklin Co., Massachusetts; 1 August 1972; T. D. Sargent (TDS).
B. C. "pretiosa" (?), ♂. Batsto, Burlington Co., New Jersey; 6 July 1974; D. F. Schweitzer (DFS).
C. C. "pretiosa," ♂. Stonington, New London Co., Connecticut; 29 June 1898; (PMNH).

Chapter V

Figure 5.3 C. *grynea* "constans," ♀. Holotype. J. Doll (USNM).

Figure 5.4 C. *cerogama* "bunkeri," ♀. Macomb Co., Michigan; 29 July 1941; P. E. Moody (Mich. State Univ.).

Figure 5.5

A. C. *lacrymosa,* ♂. Missouri; 28 August 1932; (TDS).
B C. *lacrymosa* x *palaeogama* (?), ♂. Morenci, Lenawee Co., Michigan; 1 September 1973; M. C. Nielsen (MCN).
C. C. *palaeogama,* ♂. Leverett, Franklin Co., Massachusetts; 14 August 1972; T. D. Sargent (TDS).

Figure 5.6

A. C. *amatrix,* ♂. West Hatfield, Hampshire Co., Massachusetts; 23 August 1970; C. G. Kellogg (CGK).
B. C. *amatrix* x *cara* (?), ♂. West Hatfield, Hampshire Co., Massachusetts; 23 August 1970; C. G. Kellogg (CGK).
C. C. *cara,* ♂. Leverett, Franklin Co., Massachusetts; 14 August 1969; T. D. Sargent (TDS).

Chapter VI

Figure 6.8 C. *relicta,* ♀. Leverett, Franklin Co., Massachusetts; 5 August 1968; T. D. Sargent (TDS).

Figure 6.9

(left). C. *relicta,* ♂. Leverett, Franklin Co., Massachusetts; ex ova, 1969; T. D. Sargent (TDS).
(right). C. *relicta* "clara," ♀. Leverett, Franklin Co., Massachusetts; ex ova, 1970; T. D. Sargent (TDS).

Chapter VII

Figure 7.5

(left). *C. cerogama*, ♀. Pelham, Hampshire Co., Massachusetts; 16 August 1966; T. D. Sargent (TDS).

(right). *C. cerogama* "ruperti," ♂. Otsego Co., Michigan; 13 August 1961; M. C. Nielsen (MCN).

Figure 7.6

(left). *C. relicta* "phrynia," ♂. Leverett, Franklin Co., Massachusetts; 30 July 1971; T. D. Sargent (TDS).

(right). *C. relicta* "phrynia," ♂. Near Pittsburgh, Allegheny Co., Pennsylvania; 14 August 1965; (CM).

Chapter VIII

Figure 8.5

A. *C. antinympha*, ♂. Leverett, Franklin Co., Massachusetts; 8 August 1970; T. D. Sargent (TDS).
B. *C. retecta*, ♂. Leverett, Franklin Co., Massachusetts; August 1971; T. D. Sargent (TDS).
C. *C. unijuga*, ♂. Leverett, Franklin Co., Massachusetts; 30 August 1971; T. D. Sargent (TDS).
D. *C. residua*, ♂. Fontana Dam, Graham Co., North Carolina; 8 July 1972; D. F. Schweitzer (DFS).

Figure 8.6

A. *C. ilia*, ♀. Leverett, Franklin Co., Massachusetts; 5 August 1971; T. D. Sargent (TDS).
B. *C. epione*, ♀. Leverett, Franklin Co., Massachusetts; 10 August 1971; T. D. Sargent (TDS).
C. *C. dejecta*, ♂. Leverett, Franklin Co., Massachusetts; 19 July 1971; T. D. Sargent (TDS).
D *C. flebilis*, ♂. Leverett, Franklin Co., Massachusetts; 4 September 1970; T. D. Sargent (TDS).
E. *C. retecta*, ♀. Leverett, Franklin Co., Massachusetts; 12 September 1971; T. D. Sargent (TDS).
F. *C. concumbens*, ♂. West Hatfield, Hampshire Co., Massachusetts; 17 August 1972; C. G. Kellogg (CGK).

Figure 8.7

A. *C. ultronia* "lucinda," ♂. Leverett, Franklin Co., Massachusetts; August 1971; T. D. Sargent (TDS).
B. *C. retecta,* ♂. Leverett, Franklin Co., Massachusetts; August 1971; T. D. Sargent (TDS).
C. *C. ultronia* "lucinda," ♂. Leverett, Franklin Co., Massachusetts; 11 August 1970; T. D. Sargent (TDS).
D. *C. concumbens,* ♂. Leverett, Franklin Co., Massachusetts; August 1971; T. D. Sargent (TDS).

Figure 8.8

(left). *C. ultronia* "lucinda," ♂. Leverett, Franklin Co., Massachusetts; 11 August 1971; T. D. Sargent (TDS).
(right). *C. ultronia* "nigrescens," ♀. Leverett, Franklin Co., Massachusetts; 19 August 1971; T. D. Sargent (TDS).

Figure 8.9

A. *C. epione,* ♀. Leverett, Franklin Co., Massachusetts; 31 August 1970; T. D. Sargent (TDS).
B. *C. ilia,* ♀. Leverett, Franklin Co., Massachusetts; 8 August 1971; T. D. Sargent (TDS).
C. *C. judith,* ♂. Leverett, Franklin Co., Massachusetts; 22 September 1970; T. D. Sargent (TDS).
D. *C. ilia,* ♂. Batsto, Burlington Co., New Jersey; 22 September 1973; D. F. Schweitzer (DFS).

Glossary

The terms are defined as they are used in this book. Some of them may have other meanings in other contexts.

aberration An individual with a clearly abnormal phenotype which occurs very rarely in a species. The unusual appearance of an aberration is usually the result of a freakish developmental abnormality or a rare mutation. To qualify as an aberration (rather than a form or morph), an individual must exhibit characteristics which are shown by no more than 0.01% of the members of its species. Aberrations are popularly referred to as "sports."

allele One of the alternative forms of a gene which may be present at any one particular locus on a chromosome. Every individual has two alleles, one inherited from its mother and one inherited from its father, at any one locus.

allopatric Literally, in different places, and so referring to species occupying different geographic areas.

allotype A supplemental type specimen of the opposite sex to the original (holotype).

am line Antemedial line. A usually prominent, more-or-less irregular line running across the forewing (from costa to inner margin) of most moths; located approximately one-third of the distance from the base to the outer margin of the forewing. Sometimes referred to as the transverse anterior (t.a.) line.

anomalous stimulus A stimulus unexpected to a predator on the basis of its recent experiences, resulting in physiological effects which produce responses that interfere with effective predation.

anti-hybridization device See isolating mechanism.

apostatic selection Selection of prey items by a predator which promotes inter- or intraspecific diversity (e.g., polymorphism) in that prey.

chi-square A statistical test used to determine whether an experimental result would conform to an expected result if chance alone were operating. The chi-square value is usually converted to a probability (P) value, i.e., the probability that the expected result is due to chance. When P is less than 0.05, statistical significance is claimed.

CM Carnegie Museum, Pittsburgh, Pennsylvania. Specimens from this museum were made available by Dr. Harry K. Clench.

costa The leading margin of the forewing or hindwing in butterflies and moths.

cryptic coloration The resemblance, by virtue of color and pattern, of an organism to its habitual background. This resemblance is usually enhanced by a variety of co-adapted behaviors (e.g., stillness, selection of appropriate substrates, resting attitudes, etc.).

dominant An allele whose effect is capable of masking the effect of its allelic (recessive) partner at the same locus (e.g., the allele for melanism in most moths).

form A distinctive variant within a species which occurs at a frequency greater than 0.01%, and so is characterized by genetic continuity from generation to generation, and is maintained in the species by some selective advantage. Form is here used synonymously with "morph."

founder principle The fact that a few individuals founding an outlying population of a species may, by chance, have gene frequencies which are quite different from those of their parental population.

frenulum The coupling device which holds the forewings and hindwings of a moth together during flight. A difference between males and females with regard to this coupling device (fig. 2.3) is a reliable indicator of sex in most moths.

FW Forewing(s).

gene A hereditary factor, responsible for the production of a particular enzyme or structural protein whose presence or absence may have an effect on the appearance or behavior of an individual. Genes are comprised of two alleles at a particular locus on a chromosome.

gene pool All of the genes at all of the loci in all of the individuals in a population.

genetic drift The fact that chance, rather than selection, may affect gene frequencies when a population is very small (usually fewer than fifty individuals). This phenomenon, like that of the founder principle, is due to "sampling error".

genotype The genetic constitution of an individual organism.

gynandromorph An individual in which male and female parts develop simultaneously. The resulting individual may be referred to as a sexual mosaic, i.e., showing both male and female characteristics. If, as is commonly the case, the two sides of an organism differ in sex, the individual is referred to as a bilateral gynandromorph.

heterozygous Having two different alleles of a gene at a locus on the chromosomes.

holotype The one individual from which a species (or other taxon) is described. The specimen should bear a type label, and is usually deposited in a major museum.

homozygous Having two identical alleles of a gene at a locus on the chromosomes.

HW Hindwing(s).

hybrid An individual resulting from a mating of members of two different species. Such individuals are usually sterile and/or adaptively inferior to members of both parental species, and so represent a waste of their parents' gametes. Hybrids provide evidence for a lack, or breakdown, of isolating mechanisms between the species involved.

industrial melanism The observed increase in the incidence of melanism (melanic morphs) in many species, particularly bark-like cryptic moths, in those parts of the world which are heavily industrialized.

infrasubspecific Literally, below the subspecies; and so used to describe such variants as seasonal forms, polymorphic forms, and aberrations. Infrasubspecific names have no formal taxonomic standing, but have been frequently applied and may be convenient when one wishes to refer to particular well-known variants (e.g., polymorphic forms) within a species.

instar The stage between successive molts in an insect larva (caterpillar).

intersex An individual which develops first as one sex and then the other.

interspecific Literally, between species; and so used to describe differences in appearance or behavior between or among species.

intraspecific Literally, within the species; and so used to describe differences in appearance or behavior between or among the members of a species.

isolating mechanism Some difference between species which minimizes the probability of mating between them. Isolating mechanisms include mechanical difficulties in mating (genitalic differences), and various behavioral (courtship) and ecological (habitat, seasonal) differences which tend to separate the species involved. Isolating mechanisms are sometimes referred to as anti-hybridization devices.

melanism The deposition of melanin pigment in a structure, rendering that structure blackish. In moths, generally refers to a blackening of the wings that is controlled by a single gene (as opposed to polygenic "darkening").

morph *See* form.

mosaic *See* gynandromorph and somatic mosaic.

nomenclature Anything pertaining to the formal naming of organisms, or to the system of names used for a group of organisms.

novel stimulus An unfamiliar stimulus, usually to a predator, which is avoided by reason of the startle it elicits, or because of a preference for a familiar stimulus.

P The probability that an experimental result is due to chance. The lower the value of P, the more likely it is that the result is significant, i.e., unlikely to be due to chance. Significance is usually claimed when P is 0.05 (5%) or lower.

paratype A specimen collected at the same time as the holotype, and believed to be identical with it.

phenotype The appearance and behavior of an organism; the product of its genotype and the environment in which it develops.

pheromone A chemical substance produced by one organism for purposes of communication with another member of its species.

pm line Postmedial line. A usually prominent, more-or-less irregular line running across the forewing (from costa to inner margin) of most moths; located approximately two-thirds of the distance from the base to the outer margin of the forewing. Sometimes referred to as the transverse posterior (t.p.) line.

PMNH Peabody Museum of Natural History, Yale University, New Haven, Connecticut. Specimens from this museum were made available by Dr. Charles L. Remington.

polygenic Used in reference to a trait of an organism which is influenced by a number of genes, each having small, similar and additive effects. Such traits exhibit continuous variation in a population, rather than the discontinuous variation characteristic of polymorphisms.

polymorphism The occurrence together of two or more distinctly different forms (morphs) of a species, the rarest form comprising more than 0.01% of the population and so being maintained by selection.

protean behavior Behavior which is sufficiently erratic and unsystematic to prevent its predictability to a predator, and so functioning to reduce predator efficiency.

pupa The life-cycle stage between larva and adult (imago) in insects having complete metamorphosis.

recessive An allele whose effect may be masked by its allelic (dominant) partner at the same locus. Such alleles must be present in "double dose" (homozygous) in order for their effect to be expressed in an individual.

reniform An ordinary spot on the noctuid forewing, often somewhat kidney-shaped and lying slightly within the pm line. Prominent in most *Catocala*.

Robinson trap A light-trap utilizing a mercury vapor lamp, designed by H. S. & P. J. M. Robinson on the basis of their studies of the responses of moths to light.

searching image behavior A change in the perceptual or response tendencies of a predator after a few successful encounters with a particular prey item, with the effect that the predator exhibits an increased tendency to take that particular prey item.

sex ratio The relative proportions of males and females in a sample from a population. Usually expressed as the percentage of males.

somatic mosaic An individual in which different parts have different genetic composition (though not of different sex). Such individuals usually have a bizarre appearance, and are popularly described as "freaks" or "monstrosities."

speciation The evolutionary processes leading to the formation of species.

"sport" *See* aberration.

st line Subterminal line. A line paralleling the pm line and lying approximately halfway between that line and the outer margin of the forewing. This line is often whitish, and may be blurred, vague, or obsolete in many *Catocala*.

subreniform An ordinary spot of the noctuid forewing, lying just below the reniform. The subreniform may be *open* (a continuous loop of the pm line) or *closed* (a cut-off loop from the pm line), though this difference is rarely of value in terms of specific identification.

subspecies Members of a species from a particular geographic region which are phenotypically distinct from members of the same species in other regions. Sometimes referred to as "incipient species," as it is believed that some genetic differentiation has already proceeded, and that continued isolation might produce sufficient differentiation for speciation to occur.

sugaring A technique for attracting certain moths, usually involving the application of a sugary (and slightly alcoholic) mixture to tree trunks.

sympatric Literally, in the same place; and so referring to species occupying the same geographic area.

synonym Another scientific name for a species, usually applied by an author who is unaware that the species in question has been previously named. The "rule of priority" governs in such situations, i.e., the first name published has precedence and is the name by which the species is formally known.

tapping A collecting procedure involving the rapping of tree trunks with a stick to drive off the cryptic moths. The moths may then be netted in flight, or captured when they alight again.

taxon A general term used to refer to any taxonomic category of individuals believed to be related to one another (e.g., species, genus, family, order, etc.). The plural is taxa.

taxonomy The branch of biology dealing with the naming and classifying of organisms.

type The individual(s) from which a species (or other taxon) is described. *See* allotype, holotype and paratype.

USNM United States National Museum, Smithsonian Institution, Washington D.C. Specimens from this museum were made available by Dr. Douglas C. Ferguson.

variety A general term used to refer to individuals of a species which differ from the normal in some way, including forms and aberrations.

Bibliography

Adams, M. S. & M. S. Bertoni. 1968. Continuous variation in related species of the genus *Catocala* (Noctuidae). *J. Lepid. Soc.* 22: 231–36.

Alcock, J. 1973. Cues used in searching for food by red-winged blackbirds (*Agelaius phoeniceus*). *Behavior* 46: 174–88.

Alexander, R. D. & T. E. Moore. 1962. *The Evolutionary Relationships of 17-year and 13-year Cicadas.* Misc. Publ. Mus. Zool., Univ. Michigan. No. 121. 59 pp.

Allen, J. A. & B. Clarke. 1968. Evidence for apostatic selection by wild passerines. *Nature,* Lond. 220: 501–2.

Bailey, J. S. 1877. Catocalae taken at sugar at Center, N. Y. *Canad. Entomol.* 9: 215–18.

———. 1882. Femoral tufts or pencils of hair in certain Catocalae. *Papilio* 2: 51–52.

Barnes, W. & J. McDunnough. 1918*a*. Life histories of North American species of the genus *Catocala*. *Bull. Amer. Mus. Nat. Hist.* 38: 147–77.

———. 1918*b*. *Illustrations of the North American Species of the Genus* Catocala. Mem. Amer. Mus. Nat. Hist., 3 (1). 47 pp. 22 pl.

Bartsch, R. C. B. 1916. Two new forms of Catocalae. *Lepidopterist* 1: 3.

Bauer, J. 1965. A new *Catocala* from Florida (Lepidoptera, Noctuidae). *Entomol. News* 76: 197–98.

Beutenmüller, W. 1903. Notes on some species of *Catocala*. *Bull. Amer. Mus. Nat. Hist.* 19: 505–10.

———. 1918*a*. Notes on the larvae of Catocala and their habits. *Lepidopterist* 2: 17–19.

———. 1918*b*. The food-plants of Catocala. *Lepidopterist* 2: 28–30.

———. 1918*c*. Notes on the eggs of Catocala. *Lepidopterist* 2: 33–34.

Birch, M. 1970. Pre-courtship use of abdominal brushes by the nocturnal moth, *Philogophora meticulosa* (L.) (Lepidoptera: Noctuidae). *Anim. Behav.* 18: 310–16.

Bishop, J. A. 1972. An experimental study of the cline of industrial melanism in *Biston betularia* (L.) (Lepidoptera) between urban Liverpool and rural North Wales. *J. Anim. Ecol.* 41: 209–43.

Bishop, J. A. & P. Harper. 1970. Melanism in the moth *Gonodontis bidentata*: a cline within the Merseyside conurbation. *Heredity* 25: 449–56.

Blest, A. D. 1957. The evolution of eyespot patterns in the Lepidoptera. *Behaviour* 11: 209–56.

———. 1963. Longevity, palatability and natural selection in five species of New World saturniid moth. *Nature,* Lond. 197: 1183–86.

Bowles, G. J. 1885. Catocalae — Underwing moths. *16th Ann. Rep. Entomol. Soc. Ontario*. pp. 55–60.

Brimley, C. S. 1938. *The Insects of North Carolina*. N. C. Dept. Agric., Raleigh. 560 pp.

Brower, A. E. 1922. Preparatory stages of *Catocala ulalume* Str., with larva of *C. lacrymosa* for comparison (Lepid., Noctuidae). *Entomol. News* 33: 234–36.

———. 1930*a*. An experiment in marking moths and finding them again (Lepid.: Noctuidae). *Entomol. News* 41: 10–15.

———. 1930*b*. Catocala junctura in the Ozark region. *Bull. Brooklyn Entomol. Soc.* 25: 36–38.

———. 1936. Description of a new species and a new form of *Catocala* (Lep., Noctuidae). *Bull. Brooklyn Entomol. Soc.* 31: 96–98.

———. 1947. Methods for collecting underwing moths (*Catocala*). *Lepid. News* 1: 19–20.

———. 1974. *A List of the Lepidoptera of Maine — Part 1. The Macrolepidoptera*. Life Sci. & Agric. Expt. Sta., Univ. Maine, Orono. Tech. Bull. 66. 136 pp.

Brower, L. P. 1959. Speciation in butterflies of the *Papilio glaucus* group. II. Ecological relationships and interspecific sexual behavior. *Evolution* 13: 212–28.

———. 1971. Prey coloration and predator behavior. In *Topics in the Study of Life: The BIO Source Book*. Harper & Row, New York. pp. 360–70.

Bunker, R. 1874. Notes on collecting Catocalas. *Canad. Entomol.* 6: 25–26.

Byers, C. F. 1940. A study of dragonflies of the genus *Progomphus* (*Gomphoides*). *Proc. Florida Acad. Sci.* 4: 19–86.

Carpenter, G. D. H. 1941. The relative frequency of beak-marks on butterflies of different edibility to birds. *Proc. Zool. Soc. London A* 111: 223–31.

Cassino, S. E. 1917*a*. New specics [*sic*] of Catocala. *Lepidopterist* 1: 61–64.

———. 1917*b*. A new form of Catocala ultronia. *Lepidopterist* 1: 79–80.

———. 1917*c*. A new variety of Catocala lacrymosa. *Lepidopterist* 1: 104.

———. 1918*a*. A new form of Catocala minuta. *Lepidopterist* 2: 28.

———. 1918*b*. A new form of Catocala sappho. *Lepidopterist* 2: 46–47.

———. 1918*c*. A new form of Catocala blandula Hulst. *Lepidopterist* 2: 81.

Chance, M. R. A. & W. M. S. Russell. 1959. Protean displays: a form of allaesthetic behaviour. *Proc. Zool. Soc. London A* 132: 65–70.

Clark, H. L. 1888. Preparatory stages of *Catocala relicta* Walk. *Canad. Entomol.* 20: 17–20.

Clarke, B. 1962. Balanced polymorphism and the diversity of sympatric species. *Systematics Assoc. Publ.* 4: 47–70.

———. 1964. Frequency-dependent selection for the dominance of rare polymorphic genes. *Evolution* 18: 364–69.

Clausen, L. W. 1954. *Insect Fact and Folklore*. Macmillan, New York. 194 pp.

Clench, H. K. 1947. Brief biographies. 1. William Henry Edwards. *Lepid. News* 1: 8.

Cockayne, E. A., C. N. Hawkins, F. H. Lees, B. Whitehouse & H. B. Williams. 1937–38. *Catocala fraxini*, L.: a new British record of capture and breeding. *Entomologist* 70: 241–46, 265–72; 71: 13–17, 35–38, 54–59.

Collenette, C. L. & G. Talbot. 1928. Observations on the bionomics of the Lepidoptera of Matto Grosso, Brazil. *Trans. Entomol. Soc. London* 76: 392–416.

Coppinger, R. P. 1969. The effect of experience and novelty on avian feeding behavior with reference to the evolution of warning coloration in butterflies. Part I: Reactions of wild-caught adult blue jays to novel insects. *Behaviour* 35: 45–60.

———. 1970. The effect of experience and novelty on avian feeding behavior with reference to the evolution of warning coloration in butterflies. II. Reactions of naive birds to novel insects. *Amer. Nat.* 104: 323–35.

Cott, H. B. 1940. *Adaptive Coloration in Animals*. Methuen, London. 508 pp.

Creed, E. R. 1966. Geographic variation in the two-spot ladybird in England and Wales. *Heredity* 21: 57–72.

Cross, E. W. 1896. The imitative faculty of Catocala concumbens. *Entomol. News* 7: 274.

Croze, H. J. 1970. Searching image in carrion crows. *Zeit. Tierpsychol.* 5: 1–85.

Dahm, K. H., D. Meyer, W. E. Finn, V. Reinhold & H. Roller. 1971. The olfactory and auditory mediated sex attraction in *Achroia grisella* (Fabr.). *Naturwissenschaften* 58: 265–66.

Dawkins, M. 1971. Perceptual changes in chicks: another look at the "search image" concept. *Anim. Behav.* 19: 566–74.

Dean, F. R. 1919*a*. The Catocala season of 1918 in St. Louis County, Missouri. *Lepidoptera* 3: 18–19.

———. 1919*b*. A freak Catocala palaeogama, var. phalanga. *Lepidoptera* 3: 84.

Dethier, V. G. 1963. *The Physiology of Insect Senses.* Methuen, London. 266 pp.

Dodge, E. A. 1919. Catocala notes. *Lepidoptera* 3: 54.

Dodge, G. M. 1900*a*. List of Catocalae taken at Louisiana, Missouri. *Entomol. News* 11: 433.

———. 1900*b*. Catocala Titania n. sp. *Entomol. News* 11: 472.

Downes, J. A. 1973. Lepidoptera feeding at puddle-margins, dung, and carrion. *J. Lepid. Soc.* 27: 89–99.

Driver, P. M. & D. A. Humphries. 1970. Protean displays as conflict inducers. *Nature,* Lond. 226: 968–69.

Dury, C. 1876. List of Catocalae observed in the vicinity of Cincinnati, Ohio, 1876. *Canad. Entomol.* 8: 187–88.

Dyar, H. G. 1917. Nomenclature of Catocala varieties. *Lepidopterist* 1: 31–32.

Edwards, D. K. 1962. Laboratory determinations of the daily flight times of separate sexes of some moths in naturally changing light. *Canad. J. Zool.* 40: 511–30.

Edwards, H. 1880*a*. Descriptions of some new species of Catocala. *Bull. Brooklyn Entomol. Soc.* 2: 93–97.

———. 1880*b*. Notes upon the genus Catocala, with descriptions of new varieties and species. *Bull. Brooklyn Entomol. Soc.* 3: 53–62.

Ehrmann, G. A. 1892. A local list of the genus Catocala. *Entomol. News* 3: 168–69.

———. 1894. Addition to a local list of Catocala, and a note on Papilio cresphontes. *Entomol. News* 5: 212.

———. 1918. Collecting Catocalae around the natural gas wells. *Lepidoptera* 2: 12.

Ely, C. R. 1908. Notes on C. dejecta Strecker, and other species of Catocala from East River, Connecticut. *Entomol. News* 19: 47–50.
Essig, E. O. 1931. *A History of Entomology*. Macmillan, New York. 1029 pp.
Etkin, W. 1967. *Social Behavior from Fish to Man*. Univ. Chicago Press (Phoenix), Chicago. 205 pp.

Ferguson, D. C. 1954. The Lepidoptera of Nova Scotia. Part I. Macrolepidoptera. *Proc. Nova Scotia Inst. Sci.* 23: 161–375.
Fischer, P. 1885. Description of two new varieties of Catocala cerogama, Guen., with note on a third. *Canad. Entomol.* 17: 133–34.
Forbes, W. T. M. 1954. *Lepidoptera of New York and Neighboring States. III. Noctuidae.* Cornell Univ. Agric. Expt. Sta., Mem. 329. 433 pp.
Ford, E. B. 1937. Problems of heredity in the Lepidoptera. *Biol. Rev.* 12: 461–503.
———. 1940. Genetic research in the Lepidoptera. *Ann. Eugen.*, London 10: 227–52.
———. 1964. Transient polymorphism and industrial melanism. Chap. 14 in *Ecological Genetics*. Methuen, London.
———. 1967. *Moths*. Second ed. Collins, London, 266 pp.
Foulks, O. D. 1893. Local list of Catocala. *Entomol. News* 4: 261–62.
Franclemont, J. G. 1938. Descriptions of new melanic forms (Lepidoptera: Geometridae, Noctuidae and Arctiidae). *Entomol. News* 49: 108–14.
French, G. H. 1880. Notes on Catocala hunting. *Canad. Entomol.* 12: 241–42.
———. 1881*a*. Notes on Catocala sappho Strecker. *Papilio* 1: 57.
———. 1881*b*. Some new varieties of Catocalae. *Papilio* 1: 110–11.
———. 1881*c*. A new variety of Catocala. *Papilio* 1: 218–19.
———. 1882. Preparatory stages of Catocala cara, Guen. *Papilio* 2: 167–69.
———. 1884*a*. Preparatory stages of Catocala ilia Cram. *Canad. Entomol.* 16: 12–15.
———. 1884*b*. Preparatory stages of Catocala amatrix, Hubn. *Papilio* 4: 8–10.
———. 1886. Catocala notes. *Canad. Entomol.* 18: 161–62.
———. 1892. Partial preparatory stages of Catocala illecta, Walker, with notes. *Canad. Entomol.* 24: 307–8.
———. 1894. Preparatory stages of Catocala retecta, Grote. *Canad. Entomol.* 26: 97–99.
———. 1922. *Catocala ulalume* a distinct species (Lepid., Noctuidae). *Entomol. News* 33: 233–34.

Gibb, J. A. 1962. L. Tinbergen's hypothesis of the role of specific search images. *Ibis* 104: 106–11.
Grote, A. R. 1872*a*. On the North American species of Catocala. *Trans. Amer. Entomol. Soc.* 4: 1–20.
———. 1872*b*. List of the North American species of *Catocala*. *Canad. Entomol.* 4: 164–67.

———. 1873. On the genus *Catocala*. *Canad. Entomol.* 5: 161–64.
———. 1876. On species of *Catocala*. *Canad. Entomol.* 8: 229–32.
———. 1879. A new Catocala. *Bull. Brooklyn Entomol. Soc.* 1: 77.
———. 1881*a*. Biographical sketch of M. Achille Guenée. *Papilio* 1: 31–33.
———. 1881*b*. New moths from Arizona, with remarks on Catocala and Heliothis. *Papilio* 1: 153–68.
———. 1882. Notes upon Catocala snowiana, and varieties in the genus. *Papilio* 2: 8–9.
———. 1883. Catocala concumbens, ab. hilli. *Papilio* 3: 43.
Grote, A. R. & C. T. Robinson. 1866. Lepidopterological notes and descriptions. *Proc. Entomol. Soc. Philadelphia* 6: 1–30.

Hamilton, D. W. & L. F. Steiner. 1939. Light traps and codling moth control. *J. Econ. Entomol.* 32: 867–72.
Harvey, L. F. 1877. On the black-wing group of the genus Catocala. *Canad. Entomol.* 9: 192–94.
Hiser, O. F. & J. S. Hiser. 1918. Life history of Catocala nuptialis. *Lepidopterist* 2: 66–69.
Holland, W. J. 1903. *The Moth Book*. Doubleday, New York, 479 pp. (Reprinted 1968, Dover, paperback, New York).
Hsiao, H. S. 1972. *Attraction of Moths to Light and to Infrared Radiation.* San Francisco Press, San Francisco. 89 pp.
Hulst, G. D. 1880. Remarks upon the genus Catocala, with a catalogue of species and accompanying notes. *Bull. Brooklyn Entomol. Soc.* 3: 2–13.
———. 1881. Some remarks upon the Catocalae, in reply to Mr. A. R. Grote. *Papilio* 1: 215–18.
———. 1884. The genus Catocala. *Bull. Brooklyn Entomol. Soc.* 7: 14–56.
Humphries, D. A. & P. M. Driver. 1967. Erratic display as a device against predators. *Science* 156: 1767–68.
———. 1970. Protean defence by prey animals. *Oecologia* 5: 285–302.
Hunter, M. W. III & A. C. Kamil. 1971. Object-discrimination learning set and hypothesis behavior in the northern bluejay (*Cyanocitta cristata*). *Psychon. Sci.* 22: 271–73.

Johnson, J. S. 1880. Early appearance of Catocalas. *Canad. Entomol.* 12: 137–38.
———. 1882. Catocalae taken in the vicinity of Frankford, Pennsylvania. *Canad. Entomol.* 14: 59–60.
———. 1891. Hunting Catocalae. *Entomol. News* 2: 62–65.
Jones, F. M. 1932. Insect coloration and the relative acceptability of insects to birds. *Trans Roy. Entomol. Soc. London* 82: 443–53.

Keiper, R. R. 1968. Field studies of *Catocala* behavior. *J. Res. Lepid.* 7: 113–21.
Keller, G. J. 1920. Notes on the ovum and larva of Catocala Herodias, Strecker. *Lepidopterist* 3: 121–23.

Kellicott, D. S. 1881. The larvae of Catocala flebilis and Catocala amatrix. *Papilio* 1: 141-42.

Kellogg, C. G. & T. D. Sargent. 1972. Studies on the *Catocala* (Noctuidae) of southern New England. II. Comparison of collecting procedures. *J. Lepid. Soc.* 26: 35-49.

Kettlewell, H. B. D. 1955*a*. Recognition of appropriate backgrounds by the pale and black phases of Lepidoptera. *Nature*, Lond. 175: 943-44.

———. 1955*b*. Selection experiments on industrial melanism in the Lepidoptera. *Heredity* 9: 323-42.

———. 1956. Further selection experiments on industrial melanism in the Lepidoptera. *Heredity* 10: 287-301.

———. 1957. Industrial melanism in moths and its contribution to our knowledge of evolution. *Proc. Roy. Instn.*, G. B. 36: 1-14.

———. 1958. Industrial melanism in the Lepidoptera and its contribution to our knowledge of evolution. *Proc. 10th Int. Congr. Entomol.* (1956) 2: 831-41.

———. 1973. *The Evolution of Melanism*. Clarendon, Oxford. 424 pp.

Kimball, C. P. 1965. *The Lepidoptera of Florida.* Gainesville. 363 pp.

Kirby, W. & W. Spence. 1815-26. *An Introduction to Entomology.* Longman, Rees, Orme, Brown & Green, London. 4 vols.

Klots, A. B. 1964. Notes on melanism in some Connecticut moths. *J. N. Y. Entomol. Soc.* 72: 142-44.

———. 1966. Melanism in Connecticut *Panthea furcilla* (Packard) (Lepidoptera: Noctuidae). *J. N. Y. Entomol. Soc.* 74: 95-100.

———. 1968*a*. Melanism in Connecticut *Charadra deridens* (Guenée) (Lepidoptera: Noctuidae). *J. N. Y. Entomol. Soc.* 76: 58-59.

———. 1968*b*. Further notes on melanism in Connecticut *Panthea furcilla* (Packard) (Lepidoptera: Noctuidae). *J. N. Y. Entomol. Soc.* 76: 92-95.

Koebele, A. 1881. Description of and notes upon various larvae. *Bull. Brooklyn Entomol. Soc.* 4: 20-22.

Lees, D. R., E. R. Creed & J. G. Duckett. 1973. Atmospheric pollution and industrial melanism. *Heredity* 30: 227-32.

Lemmer, F. 1937. New Lepidoptera from the New Jersey pine barrens. *Bull. Brooklyn Entomol. Soc.* 32: 22-25.

Linsley, E. G. & J. W. McSwain. 1958. The significance of floral constancy among bees of the genus *Diadasia* (Hymenoptera, Anthophoridae). *Evolution* 12: 219-23.

McDunnough, J. 1938. *Check List of the Lepidoptera of Canada and the United States of America. Part 1. Macrolepidoptera.* Mem. So. Calif. Acad. Sci. Vol. 1. 275 pp.

Mackintosh, N. J. 1965. Selective attention in animal discrimination learning. *Psychol. Bull.* 64: 124-50.

Marshall, G. A. K. & E. B. Poulton. 1902. Five years' observations and experiments (1896-1901) on the bionomics of South African insects, chiefly directed to the investigation of mimicry and warning colours. *Trans. Entomol. Soc. London*: 287-584. 15 pls.

Masters, J. H. 1972. A proposal for the uniform treatment of infrasubspecific variation by lepidopterists. *J. Lepid. Soc.* 26: 249-60.

Mather, K. 1955. Polymorphism as an outcome of disruptive selection. *Evolution* 9: 52-61.

Mayfield, T. D. 1922. Notes on the life histories of North American Catocalae, with description of two new forms. *Bull. Brooklyn Entomol. Soc.* 17: 114-20, 138-42.

———. 1923. A new form of Catocala gracilis Edwards. *Bull. Brooklyn Entomol. Soc.* 18: 33.

Mayr, E. 1963. *Animal Species and Evolution*. Harvard Univ. Press, Cambridge. 797 pp.

Merrifield, F. 1890. Systematic temperature experiments on some Lepidoptera in all their stages. *Trans. Entomol. Soc. London*, 131-59.

———. 1891. Conspicuous effects on the markings and colouring of Lepidoptera caused by exposure of the pupae to different temperature conditions. *Trans. Entomol. Soc. London*, 155-67.

Meyer, J. H. 1952. Ein neuer *Catocala*-hybrid. *Zeit. Wiener Entomol. Gesell.* 37: 65-71.

Monk, J. H., L. J. Monk & H. S. Heikens. 1960. Further evidence for the role of "searching image" in the hunting behavior of titmice. *Arch. Neerl. Zool.* 13: 448-65.

Muller, J. 1960. A new melanic form of *Catocala connubialis* from New Jersey (Noctuidae). *J. Lepid. Soc.* 14: 177-78.

———. 1973. Second addition to the supplemental list of macrolepidoptera of New Jersey. *J. N. Y. Entomol. Soc.* 81: 66-71.

Murray, W. 1877. On capturing Catocalas in the day-time. *Canad. Entomol.* 9: 18-19.

Murton, R. K. 1971. The significance of a specific search image in the feeding behaviour of the wood pigeon. *Behaviour* 40: 10-42.

Nabokov, V. 1947. (Letter to C. L. Remington). *Lepid. News* 1: 34.

Newman, E. 1835. *The Grammar of Entomology*. Westley & Davis, London. 304 pp.

Norris, M. J. 1936. The feeding habits of the adult Lepidoptera Heteroneura. *Trans. Roy. Entomol. Soc. London* 85: 61-90.

Oldroyd, H. 1958. *Collecting, Preserving and Studying Insects*. Hutchinson, London. 327 pp.

Owen, D. F. 1961. Industrial melanism in North American moths. *Amer. Nat.* 95: 227-33.

———. 1962. The evolution of melanism in six species of North American geometrid moths. *Ann. Entomol. Soc. Amer.* 55: 695-702.

Owen, D. F. & M. S. Adams. 1963. The evolution of melanism in a population of *Catocala ilia* (Noctuidae). *J. Lepid. Soc.* 17: 159-62.

Payne, J. A. & E. W. King. 1969. Lepidoptera associated with pig carrion. *J. Lepid. Soc.* 23: 191-95.

Peattie, D. C. 1935. *An Almanac For Moderns*. Putnam, New York, 396 pp.
———. 1938. *Green Laurels: The Lives and Achievements of the Great Naturalists*. Garden City Publ. Co., New York. 368 pp.
Pilate, G. R. 1882. A new variety of Catocala. *Papilio* 2: 31-32.
Platt, A. 1969. A lightweight collapsible bait trap for Lepidoptera. *J. Lepid. Soc.* 23: 97-101.
Poulton, E. B. 1887. The experimental proof of the protective value of colour and markings in insects in reference to their vertebrate enemies. *Proc. Zool Soc. London*, 191-274.
———. 1890. *The Colours of Animals*. Appleton, New York, 360 pp.
———. 1913. Disabling and other injuries found in the Lepidoptera and their interpretation. *Proc. Entomol. Soc. Lond.*, xix-xxii.

Rabinowitch, V. 1968. The role of experience in the development of food preferences in gull chicks. *Anim. Behav.* 16: 425-28.
———. 1969. The role of experience in the development and retention of seed preferences in zebra finches. *Behaviour* 33: 222-36.
Rau, P. & N. L. Rau. 1929. The sex attraction and rhythmic periodicity in giant saturniid moths. *Trans. Acad. Sci. St. Louis* 26: 83-221.
Reiff, W. 1916. Catocala amica Hb. subspecies novangliae Reiff. *Lepidopterist* 1: 12-15.
———. 1918*a*. The foodplants of the North American species of the genus Catocala. Lepidoptera — Heterocera. *Lepidoptera* 2: 27-28.
———. 1918*b*. A new form of Catocala minuta. *Lepidoptera* 2: 46.
———. 1919. Catocala herodias Strecker. *Lepidoptera* 3: 73-74.
———. 1919-1920. Notes and additions to Barnes' and McDunnough's Illustrations of the N. A. Species of the Genus Catocala. *Lepidoptera* 3: 69-70, 75-76, 86-87, 92-94; 4: 12-13, 21-22, 39-40, 46-48, 54-56, 62-64.
Remington, C. L. 1954. The genetics of *Colias* (Lepidoptera). *Adv. Genet.* 6: 403-50.
———. 1958. Genetics of populations of Lepidoptera. *Proc. 10th Int. Congr. Entomol.* (1956) 2: 787-805.
Remington, J. E. 1948*a*. Brief biographies. 9. Henry Edwards (1830-1891). *Lepid. News* 2: 7.
———. 1948*b*. Brief biographies. 10. Augustus Radcliffe Grote (1841-1903). *Lepid. News* 2: 17.
Riddiford, L. M. & C. M. Williams. 1967*a*. Volatile principle from oak leaves: role in sex life of the polyphemus moth. *Science* 155: 589-90.
———. 1967*b*. Chemical signaling between polyphemus moths and between moths and host plant. *Science* 156: 541.
Robinson, H. S. 1952. On the behaviour of night-flying insects in the neighbourhood of a bright source of light. *Proc. Roy. Entomol. Soc. London A* 27: 13-21.
Robinson, H. S. & P. J. M. Robinson. 1950. Some notes on the observed behavior of Lepidoptera in the vicinity of light sources together with a description of a light-trap designed to take entomological samples. *Entomol. Gaz.* 1: 3-20.

Roeder, J. K. 1966. Auditory system of noctuid moths. *Science* 154: 1515–21.
Rowley, R. R. 1908. Notes on Catocala. *Entomol. News* 19: 115–20.
———. 1909. Another season with Catocalae. *Entomol. News* 20: 127–35.
Rowley, R. R. & L. Berry. 1909. Notes on the study of some Iowa Catocalae. *Entomol. News* 20: 12–18.
———. 1910*a*. Further study of the Catocalae. *Entomol. News* 21: 104–16.
———. 1910*b*. Notes on the life stages of Catocalae; a summer's record and incidental mention of other lepidoptera. *Entomol. News* 21: 441–55.
———. 1912. A dry year's yield of Catocalae (Lepid.) 1911. *Entomol. News* 23: 207–14.
———. 1913. Last year's work with Catocalae and other Lepidoptera. *Entomol. News* 24: 197–205.
———. 1914. 1913 as a Catocala year (Lepid.). *Entomol. News* 25: 157–67.
Royama, T. 1970. Factors governing the hunting behaviour and selection of food by the great tit. *J. Anim. Ecol.* 39: 619–68.
Ruiter, L. de. 1952. Some experiments on the camouflage of stick caterpillars. *Behaviour* 4: 222–32.

Saario, C. A., H. H. Shorey & L. K. Gaston. 1970. Sex pheromones of noctuid moths. XIX. Effect of environmental and seasonal factors on captures of males of *Trichoplusia ni* in pheromone-baited traps. *Ann. Entomol. Soc. Amer.* 63: 667–72.
Sargent, T. D. 1966. Background selections of geometrid and noctuid moths. *Science* 154: 1674–75.
———. 1968. Cryptic moths: effects on background selections of painting the circumocular scales. *Science* 159: 100–101.
———. 1969*a*. Behavioral adaptations of cryptic moths. II. Experimental studies on bark-like species. *J. N. Y. Entomol. Soc.* 77: 75–79.
———. 1969*b*. Behavioral adaptations of cryptic moths. III. Resting attitudes of two bark-like species, *Melanolophia canadaria* and *Catocala ultronia*. *Anim. Behav.* 17: 670–72.
———. 1969*c*. A suggestion regarding hindwing diversity among moths of the genus *Catocala* (Noctuidae). *J. Lepid. Soc.* 23: 261–64.
———. 1969*d*. Background selections of the pale and melanic forms of the cryptic moth, *Phigalia titea* (Cramer). *Nature*, Lond. 222: 585–86.
———. 1971. Melanism in *Phigalia titea* (Cramer) (Lepidoptera: Geometridae). *J. N. Y. Entomol. Soc.* 79: 122–29.
———. 1972*a*. Studies on the *Catocala* (Noctuidae) of southern New England. III. Mating results with *C. relicta* Walker. *J. Lepid. Soc.* 26: 94–104.
———. 1972*b*. Sketches of New England moths. 5. Polymorphisms. *Man & Nature* (September): 25.
———. 1973*a*. Behavioral adaptations of cryptic moths. VI. Further experimental studies on bark-like species. *J. Lepid. Soc.* 27: 8–12.
———. 1973*b*. Studies on the *Catocala* (Noctuidae) of southern New Eng-

land. IV. A preliminary analysis of beak-damaged specimens, with discussion of anomaly as a potential anti-predator function of hindwing diversity. *J. Lepid. Soc.* 27: 175–92.

———. 1974. Melanism in moths of central Massachusetts (Noctuidae, Geometridae). *J. Lepid. Soc.* 28: 145–52.

Sargent, T. D. & S. A. Hessel. 1970. Studies on the *Catocala* (Noctuidae) of southern New England. I. Abundance and seasonal occurrence of the species, 1961–1969. *J. Lepid. Soc.* 24: 105–17.

Sargent, T. D. & R. R. Keiper. 1969. Behavioral adaptations of cryptic moths. I. Preliminary studies on bark-like species. *J. Lepid. Soc.* 23: 1–9.

Sargent, T. D. & D. F. Owen. 1975. Apparent stability in hindwing diversity in samples of moths of varying species composition. *Oikos* 26: 205–10.

Schneider, D. 1962. Electrophysiological investigation on the olfactory specificity of sexual attracting substances in different species of moths. *J. Insect Physiol.* 8: 15–30.

Schwarz, E. 1915. Recent work on Catocalae: a new aberration and correction (Lep.). *Entomol. News* 26: 289–90.

———. 1916. Observations on the habits of Catocala titania Dodge (Lepid.) *Entomol. News* 27: 67–69.

———. 1919. On the early stages of Catocala titania Dodge, and a description of three new varieties of Catocala (Lep.). *Entomol. News* 30: 14–17.

Seitz, A. 1914. The Macrolepidoptera of the World. Section I. Vol. 3. *The Palearctic Noctuidae* (text and plates). Stuttgart.

Sevastopulo, D. G. 1974. A proposal for the uniform treatment of infrasubspecific variation by lepidopterists. *J. Lepid. Soc.* 28: 289–90.

Shorey, H. H. 1964. Sex pheromones of noctuid moths. II. Mating behavior of *Trichoplusia ni* (Lepidoptera: Noctuidae) with special reference to the role of the sex pheromone. *Ann. Entomol. Soc. Amer.* 57: 371–77.

———. 1966. The biology of *Trichoplusia ni* (Lepidoptera: Noctuidae). IV. Environmental control of mating. *Ann. Entomol. Soc. Amer.* 59: 502–6.

Shorey, H. H. & L. K. Gaston. 1965*a*. Sex pheromones of noctuid moths. V. Circadian rhythm of pheromone-responsiveness in males of *Autographa californica*, *Heliothis virescens*, *Spodoptera exigua*, and *Trichoplusia ni* (Lepidoptera: Noctuidae). *Ann. Entomol. Soc. Amer.* 58: 597–600.

———. 1965*b*. Sex pheromones of noctuid moths. VII. Quantitative aspects of the production and release of pheromone by females of *Trichoplusia ni* (Lepidoptera: Noctuidae). *Ann. Entomol. Soc. Amer.* 58: 604–8.

———. 1970. Sex pheromones of noctuid moths. XX. Short-range orientation by pheromone-stimulated males of *Trichoplusia ni*. *Ann. Entomol. Soc. Amer.* 63: 829–32.

Shorey, H. H., L. K. Gaston & J. S. Roberts. 1965. Sex pheromones of noctuid moths. VI. Absence of behavioral specificity for the female sex pheromones of *Trichoplusia ni* versus *Autographa californica* and *Heliothis zea* versus *H. virescens* (Lepidoptera: Noctuidae). *Ann. Entomol. Soc. Amer.* 58: 600-603.

Shorey, H. H., S. U. McFarland & L. K. Gaston. 1968. Sex pheromones of noctuid moths. XIII. Changes in pheromone quantity, as related to reproductive age and mating history, in females of seven species of Noctuidae (Lepidoptera). *Ann. Entomol. Soc. Amer.* 61: 372-76.

Shorey, H. H., K. L. Morin & L. K. Gaston. 1968. Sex pheromones of noctuid moths. XV. Timing of development of pheromone-responsiveness and other indicators of reproductive age in males of eight species. *Ann. Entomol. Soc. Amer.* 61: 857-61.

Sillman, A. J. 1973. Avian vision. In *Avian Biology*, D. S. Farner & J. R. King, eds. Vol. 3. Academic Press, New York, pp. 349-87.

Smyth, E. A., Jr. 1899. The Catocalae of Montgomery County, Virginia. *Entomol. News* 10: 282-86.

Snyder, A. J. 1897*a*. A remarkable appearance of Catocala insolabilis. *Canad. Entomol.* 29: 76.

———. 1897*b*. A rare Catocala. *Canad. Entomol.* 29:220.

Sokolov, E. N. 1960. Neuronal models and the orienting reflex. In *The Central Nervous System and Behavior*, M. A. B. Brazier, ed. Joseph Macy, Jr. Found., New York. pp. 187-276.

Sower, L. L., H. H. Shorey & L. K. Gaston. 1970. Sex pheromones of noctuid moths. XXI. Light-dark cycle regulation and light inhibition of sex pheromone release by females of *Trichoplusia ni*. *Ann. Entomol. Soc. Amer.* 63: 1090-92.

———. 1971. Sex pheromones of noctuid moths. XXV. Effects of temperature and photoperiod on circadian rhythms of sex pheromone release by females of *Trichoplusia ni*. *Ann. Entomol. Soc. Amer.* 64: 488-92.

Spieth, H. T. 1952. Mating behavior within the genus *Drosophila* (Diptera). *Bull. Amer. Mus. Nat. Hist.* 99: 395-474.

Stowers, N. 1916. A sugaring trip for Catocalas. *Lepidopterist* 1: 87-89.

Strecker, H. 1872-77. *Lepidoptera, Rhopaloceres and Heteroceres, Indigenous and Exotic, with Descriptions and Colored Illustrations.* Reading, Pennsylvania. 143 pp.

———. 1897. Catocala jair — new species from Florida. *Entomol. News* 8: 116-17.

Swynnerton, C. F. M. 1926. An investigation into the defences of butterflies of the genus *Charaxes*. *Proc. 3rd Intern. Entomol. Congr.*, Zurich (1925) 2: 478-506.

Tietz, H. M. 1972. *An Index to the Described Life Histories, Early Stages and Hosts of the Macrolepidoptera of the Continental United States and Canada.* 2 vols. Allyn Mus. Entomol., Sarasota, Florida.

Tinbergen, L. 1960. The natural control of insects in pine woods. I. Factors influencing the intensity of predation by song birds. *Arch. Neerl. Zool.* 13: 265-343.

Turner, E. R. A. 1961. Survival value of different methods of camouflage as shown in a model population. *Proc. Zool. Soc. London* 136: 273–83.

Urquhart, F. A. 1960. *The Monarch Butterfly*. Univ. Toronto Press. 361 pp.

Villiard, P. 1969. *Moths and How to Rear Them*. Funk & Wagnalls. New York. 242 pp.

Walker, F. 1854–66. *List of Specimens of Lepidopterous Insects in the Collection of the British Museum*. 35 pts. London.

Wigglesworth, V. B. 1972. *The Principles of Insect Physiology*. Seventh ed. Chapman & Hall, London. 827 pp.

Wilkinson, R. S. 1971*a*. Bio-bibliographical foreword. In *An Illustrated Essay on the Noctuidae of North America, with A Colony of Butterflies*, A. R. Grote 1882; reprinted by E. W. Classey, Ltd., Middlesex, England.

———. 1971*b*. Daylight collecting of *Catocala* (Lepidoptera; Noctuidae). *Michigan Entomol.* 4: 59–60.

Williams, C. B. 1935. The times of activity of certain nocturnal insects, chiefly Lepidoptera, as indicated by a light-trap. *Trans. Roy. Entomol. Soc. London* 83: 523–55.

———. 1939. An analysis of four years captures of insects in a light trap. Part 1. General survey; sex proportion phenology; and time of flight. *Trans. Roy. Entomol. Soc. London* 89: 79–131.

———. 1958. *Insect Migration*. Collins, London. 235 pp.

Wilson, E. O. & W. H. Bossert. 1963. Chemical communication among animals. *Recent Prog. Hormone Res.* 19: 673–716.

Wormsbacher, H. 1918. Collecting Catocalae along Lake Erie. *Lepidoptera* 2: 54.

Worthington, C. E. 1883. On certain Catocala. *Papilio* 3: 39–41.

Wyman, L. C. & F. L. Bailey. 1964. *Navaho Indian Ethnoentomology*. Univ. New Mexico Press. Alburquerque. 158 pp.

Young, F. N. 1958. Some facts and theories about the broods and periodicity of the periodical cicadas. *Proc. Indiana Acad. Sci.* 68: 164–70.

Index

References to illustrations are in italic; those to species accounts are in boldface.